Analysis of Radome-Enclosed Antennas

Dennis J. Kozakoff

Artech House
Boston • London

Library of Congress Cataloging-in-Publication Data
Kozakoff, D. J.
Analysis of radome-enclosed antennas / Dennis J. Kozakoff.
p. cm.
Includes bibliographical references and index.
ISBN 0-89006-716-3 (alk. paper)
1. Radomes. I. Title.
TK6590.R3K69 1997
621.3848'3—dc21 97-27103
CIP

British Library Cataloguing in Publication Data
Kozakoff, Dennis J.
Analysis of radome-enclosed antennas
1. Radomes
I. Title
621.3'8483

ISBN 0-89006-716-3

Cover design by Eileen Hoff

685 Canton Street
Norwood, MA 02062

International Standard Book Number: 0-89006-716-3
Library of Congress Catalog Card Number: 97-27103

10 9 8 7 6 5 4 3 2 1

Analysis of Radome-Enclosed Antennas

Contents

Preface	xi
Objectives	xi
Skills and Hardware You Will Need	xii
What You Can Expect From Radome Analysis	xii
Organization of the Book	xiii
Acknowledgments	xv
Part I Background and Fundamentals	1
Chapter 1 Overview of Radome Analysis	3
1.1 What Is a Radome?	3
1.2 History of Radome Development	3
1.3 Radome–Antenna Interaction	5
Boresight Error and Boresight Error Slope	6
Increased Antenna Sidelobe Levels	7
Depolarization	7
Insertion Loss as a Result of the Presence of a Radome	7
1.4 Significant Parameters in Radome Analysis	8
References	10
Chapter 2 Basic Principles and Conventions	11
2.1 Vector Mathematics	11
2.2 Electromagnetic Theory	14
2.3 Matrices	17
2.4 Coordinate System and Gimbal Relationships	19
References	24

Chapter 3 Antenna Fundamentals 27
3.1 Directivity and Gain 27
3.2 Radiation From Current Elements 30
3.3 Antenna Array Factor 31
3.4 Linear Aperture Distributions 32
3.5 Two-Dimensional Distributions 37
3.6 Spiral Antennas 41
References 43
Selected Bibliography 45

Chapter 4 Radome Dielectric Materials 47
4.1 Organic Radome Dielectric Wall Materials 47
Sandwich Wall Radomes 48
Core Materials 50
4.2 Inorganic Radome Dielectric Wall Materials 53
4.3 Dual-Mode (RF/IR) Radome Wall Materials 56
Nonorganic Dual-Mode Materials 57
Organic Dual-Mode Materials 59
4.4 Effect of Radome Material on Antenna Noise Temperature 60
References 63

Part II Radome Analytical Techniques 67

Chapter 5 Multilayer Dielectric Walls 69
5.1 Mathematical Formulation for Radome Wall Transmission 69
Transmission Coefficients for Circular Polarized Waves 72
5.2 Radome Types, Classes, and Style Definitions 72
5.3 Wall Style Electrical Performance 73
Half-Wave Wall Radomes (Style a) 75
Thin-Walled Radomes (Style b) 78
A-Sandwich Radomes (Style c) 78
Multilayered Wall Radomes (Style d) 79
B-Sandwich Walled Radomes (Style e) 81
References 85
Selected Bibliography 85
Appendix 5A: Program Wall Computer Software Listing 85

Chapter 6 Radome Analysis Techniques 97
6.1 Background 97
6.2 Geometric Optics Approaches to Propagation Through Radomes 100
Geometrical Optics Receiving Mode Calculations 102
Geometrical Optics Transmitting Mode Calculations 104

6.3 Physical Optics Approaches to Propagation Through Radome Walls 105
Physical Optics Receiving Mode Calculations 105
Physical Optics Transmitting Mode Calculations 107
6.4 Other Techniques 108
Method of Moments 108
Plane Wave Spectra 108
6.5 Sources of Ray-Trace Computational Error 109
Internal Reflections 110
Wall Model and Statistical Variations 114
References 114

Part III Computer Implementation 117

Chapter 7 Radome Modeling Approaches 119
7.1 Tangent Ogive Radome Shape 120
Rationale for Choosing a Particular Shape 120
Mathematical Modeling of Shape 123
Computation of Surface Normal Vector 125
7.2 Determining the Intercept Point of a Ray with a Tangent Ogive Radome Wall 127
Method 1 127
Method 2 127
Method 3 128
7.3 Intercept of Ray With General Radome Shapes Having Axial Symmetry 134
Basic Information About the POLY Program 135
Determining Intercept Point and Surface Normal 136
7.4 Wave Decomposition at Intercept Point 136
References 139
Appendix 7A: Program BRUTE Software Listing 140
Appendix 7B: Program ROOTS Software Listing 142
Appendix 7C: Program POLY Software Listing 145
Appendix 7D: Program COMPOSE Software Listing 149

Chapter 8 Radome-Enclosed Monopulse Antennas 153
8.1 Definition of Boresight Error and Boresight Error Slope 155
8.2 Calculation of Antenna Patterns and Monopulse Error Voltages for an Antenna Without a Radome 158
8.3 Calculation of Antenna Patterns and Monopulse Error Voltages for a Radome-Enclosed Antenna 169
Step 1: Compute the Antenna Pattern Space Phase Terms 170
Step 2: Transpose All Aperture Points to Radome Coordinates 170

Step 3: Transform All Ray Vectors from Aperture Sample Points in the Pattern Look Direction 172
Step 4: Project Rays to Radome Surface 172
Step 5: Angle-of-Incidence Determination 172
Step 6: Compute Complex Voltage Transmission Coefficients for Each Ray 172
Step 7: Perform an Antenna Aperture Integration 173
8.4 Additional Radome Modeling Considerations 173
Rain Erosion of Radome 175
Aerothermal Heating 176
Computation of Radome Effects on Conscan Antennas 177
References 177
Selected Bibliography 179
Appendix 8A: Program PATTERN Software Listing 179

Chapter 9 Other Radome-Enclosed Antenna Types 183
9.1 Spiral Antennas 183
Multimode Spiral Antennas 188
9.2 Parabolas and Other Telecommunications Antennas 190
9.3 Air-Supported Radomes 192
9.4 Metal Space Frame Radomes 192
Dielectric Wall Losses 194
Frame Blockage Losses 195
Pointing Errors 196
Dielectric Space Frame Radomes 196
9.5 Radome-Enclosed, Phased-Array Antennas 198
References 202
Selected Bibliography 203

Appendix A Vector Operations in Various Coordinate Systems 205
Rectangular Coordinates 205
Cylindrical Coordinates 205
Spherical Coordinates 206

Appendix B Propagation Constant and Wave Impedance for Arbitrary Media 207
Wave Components in Media 207
Derivation of Propagation Constant and Apparent Permittivity 210
Wave Impedance 211

Appendix C Radiation From Elemental Currents 213

Appendix D Listing of Program Tangent Ogive Radome 221
Variable List 221
Monopulse 227
Subroutine Monopulse Antenna 227

Appendix E Operations Manual for Tangent Ogive Radome Software 239
Installation 239
System Requirements for Version 1.0 240
Input/Output Requirements 240
Supported Printers 240
E.1 Executing the Program 240
E.2 Flat Panel Analysis (Wall.Exe) 240
E.2.1 Theory 240
E.2.2 Operation 241
E.3 Monopulse Antenna Analysis 241
E.3.1 Theory 241
E.3.2 Operation 242

Glossary 245

List of Symbols and Abbreviations 251
Greek 254

About the Author 257

Index 259

Preface

If you are reading this book, you probably fit one of the following occupational profiles:

- An engineer, physicist, or scientist with a career interest in antennas and radomes;
- A program manager in government or industry with the job of buying radomes or evaluating radomes that have already been installed.

If one of these occupational profiles describes you, this book has been targeted especially for your needs. If you are just reading this book for general interest, you too should be able to learn about radomes.

OBJECTIVES

The ultimate goal of this book is to help you develop your own advanced software for radome analysis. To accomplish this goal, you will be led step by step through the building blocks and individual problems that must be solved on the way to the ultimate goal.

If you read and study this book thoroughly, you should be able to:

- Define a radome, describe how it works, tell the history of its development, and understand the parameters determining a good approach to analysis;
- Develop the framework of mathematical and electromagnetic theory needed for analyzing the effects of a radome on an antenna;
- Understand the basic fundamentals of antennas;
- Understand how radome materials and composition affect specific analytical techniques;

- Choose the best radome modeling technique for your particular antenna;
- Understand and use the software accompanying this book.

SKILLS AND HARDWARE YOU WILL NEED

Although radome analysis does not require you to have the math skills of a Ph.D. scientist, you do need to understand the fundamental equations presented in this book.

In addition, you need programming skills and a knowledge of the BASIC language. This language is easy to learn, and you do not need to be a high-level programmer to work with radome analysis.

The software accompanying this book is programmed in BASIC. Because the BASIC on the diskette is in compiled form, the software should run at about the same speed as a PASCAL or C program. All software coding in the book is given in BASIC.

You will need the following hardware to run the programs:

- A minimum of a 386 IBM-compatible computer with math coprocessor;
- 4 MB of RAM;
- An Epson-compatible printer to output data.

If you have a Pentium, the program will work more efficiently.

WHAT YOU CAN EXPECT FROM RADOME ANALYSIS

Radome analysis is something of an art as well as a science. You cannot guarantee that your solution to a given problem will be exact. The approach to analysis taught in this book uses many approximations. However, you can usually obtain a good ballpark estimate of radome performance.

In this book, the approaches to problem solving and the examples shown are believed to be correct by the author and are believed to get the user to the correct ballpark solution. The results of any problem demonstrated in this book may vary somewhat from case to case.

Radome analysis consists of dividing the overall problem into many smaller problems. This method of analysis is reinforced by the choice of the BASIC programming language, which supports the following data types: integers, reals, booleans, floating point, strings, arrays, and frames. (A *frame* is a list of field names, each with a value that can be any of the allowable data types.) Compared to higher order languages such as Fortran, BASIC is deficient in that it cannot handle complex numbers. Thus, in programming electromagnetic mathematics in this book, it was necessary to divide the problem into real and imaginary parts and then do operations on each part independently.

ORGANIZATION OF THE BOOK

This book contains three sections:

- Part I covers the elementary background material on radomes, the relevant mathematical and electromagnetic theory, and the fundamentals of antennas.
- Part II is an overview of radome modeling techniques.
- Part III discusses software programming approaches to achieve the modeling on a PC computer.

Acknowledgments

This book could not have been completed without the help of my wife, E. Ruth Kozakoff, who has typed the original manuscript and been a source of constant encouragement. I am also thankful to the many others who have contributed to this effort. Dr. Don G. Bodnar, chief scientist of the U.S.A.F. Rome Laboratory, and Dr. Donald J. Grace, emeritus vice president and director of the Georgia Tech Research Institute, have reviewed the manuscript and provided valuable comments and direction to this project. Ben Mackenzie, principal engineer at Norton Company Composites Division, has provided useful data regarding aircraft radome materials. Rich Matyskiela, system engineer at AEL Industries, has contributed unique insights in antenna gimbal transformations. Dennis Kozakoff, Jr., of Millimeter Wave Technology, Inc., and Pat Watkins of Electromagnetic Sciences, Inc., have contributed in the coding of computer software. Finally, J. Robert Pirkle of Millimeter Wave Technology, Inc., has provided Autocad artwork for the book.

PART I

Background and Fundamentals

Chapter 1

Overview of Radome Analysis

1.1 WHAT IS A RADOME?

A *radome,* an acronym coined from *radar dome,* is a structure placed over an antenna that protects the antenna from its physical environment. Ideally, the radome is radio-frequency (RF) transparent and does not degrade the electrical performance of the enclosed antenna.

Today, radomes find wide applications in ground, maritime, aircraft, and missile electronic systems. For instance, Figure 1.1(a) is a close-up view of a radome-covered satellite communications (SATCOM) antenna. Figure 1.1(b) shows the radome installed on a small boat.

1.2 HISTORY OF RADOME DEVELOPMENT

Before the radome, slow-speed aircraft in early World War II carried VHF radars that used externally mounted Yagi or dipole-array antennas. With the advent of microwave radar on higher speed aircraft, radomes became necessary.

Airborne microwave radar appeared in the United States in 1940, when Sperry Rand Corporation tested an experimental unit, suspended with no radome in the open bomb bay of a B-18 aircraft [1]. At about the same time, the introduction of the microwave magnetron from England spawned the rapid development of both military and civilian radars.

The first reported aircraft radomes used simple, thin-wall designs. In 1941, the first in-flight radome was a hemispherical nose radome fabricated from plexiglass. It protected an experimental S-band, Western Electric radar flown in a B-18A aircraft. Beginning in 1943, production airborne radars used plywood radomes that were 0.25 in. thick [2]. In this period, plywood radomes also appeared on Navy PT boats and blimps, as well as in ground installations.

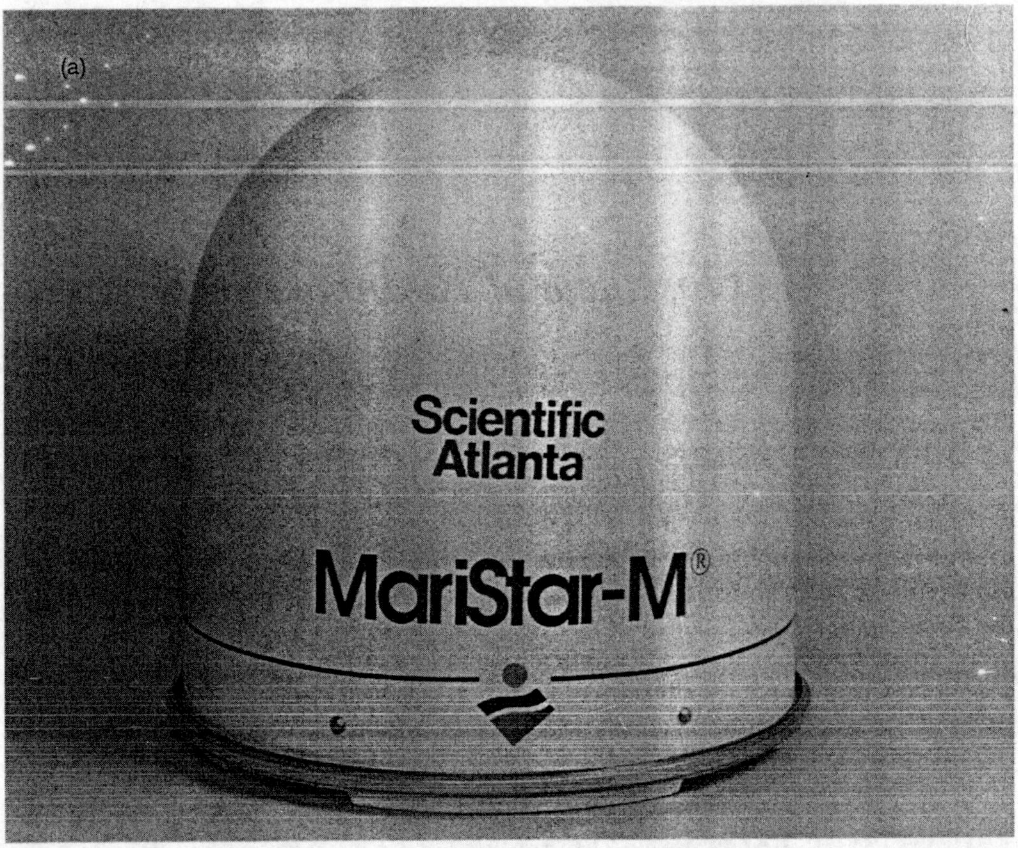

Figure 1.1 Radome-enclosed satellite communications antenna: (a) close-up view of a radome and (b) marine installation of a radome. (Photographs courtesy of Scientific-Atlanta Inc.)

Because plywood has moisture absorption problems and does not easily bend into doubly curved shapes, new radome construction techniques and materials were necessary. For instance, in 1944 the MIT Radiation Laboratory developed the three-layer "A-sandwich," which used dense skins and a low-density core material. Common during this era were fiberglass skins and polystyrene fiber core materials [3].

Since World War II, radome materials have developed in two main areas:

1. Ceramics, used primarily for high-speed missile radomes;
2. High-strength organic materials for sandwich composite radomes.

Today, the majority of aircraft radomes use sandwich-wall construction. Figure 1.2 shows an airplane with a weather-radar nose radome designed with sandwich wall construction.

Figure 1.1 (continued)

Advanced computerized design analysis techniques have also appeared as the result of electronic performance requirements for new avionics and radars. The distribution of radome technology data through the various symposiums is the primary force advancing computerized design analysis techniques. Today, the most well-known of these symposiums is probably the biennial Electromagnetic Window Symposium held at the Georgia Institute of Technology in Atlanta, Georgia.

1.3 RADOME–ANTENNA INTERACTION

During recent decades, rapid growth in the use of the electromagnetic spectrum has led to many improvements in the performance of electromagnetic sensors. Two examples of these improvements are listed:

1. Dual-polarized, wideband antenna systems for improved tracking and radar performance;

Figure 1.2 Weather-radar radome on nose of C-130 aircraft. (Photo courtesy of Lockheed Martin Aeronautical Systems.)

2. Very low sidelobe antennas that provide ground clutter rejection for airborne radars that use pulse-Doppler modes.

These sensors require more stringent radome performance than their predecessors because the antenna performance can be altered by radome effects on its radiation patterns. These effects include, but are not limited to, the following:

- Boresight error (BSE) and boresight error slope (BSES);
- Increased antenna sidelobe levels;
- Depolarization;
- Introduction of an insertion loss due to the presence of the radome.

Boresight Error and Boresight Error Slope

Boresight error is a bending of the angle-of-arrival of a received signal relative to its actual angle-of-arrival. This phenomenon stems primarily from distortions of the electromagnetic wavefront as it propagates through a dielectric radome wall. BSES is defined as the rate of change of BSE as a function of the antenna scan angle within the radome.

For a radome-enclosed missile monopulse antenna, BSE depends on frequency, polarization, and antenna orientation. Radome BSE directly affects the "miss distance" of a missile flying a pursuit guidance trajectory. Radome BSES also impacts the miss distance of a missile flying a proportional or modern guidance law trajectory [4].

Radome BSES can cause severe degradation for modern guidance systems as well as for classical proportional navigation systems. Missile guidance system designers often use computer and/or hardware-in-the-loop (HWIL) simulations to determine the relationship between a missile terminal miss distance (or guidance accuracy) versus the BSES of the radome. The desired guidance accuracy will dictate a maximum allowable BSES.

In production, the radome manufacturing tolerance affects the BSES characteristics of the radome. As a result of the computer and/or HWIL simulations just discussed, the maximum allowable BSES is often used to specify the required manufacturing tolerances of the radome [5].

Increased Antenna Sidelobe Levels

Antenna sidelobes generally increase when an antenna is placed within a radome. This increase occurs because of wavefront distortion and wall transmission effects as a wavefront propagates through a radome wall. The state of the art of low-sidelobe-level antennas has progressed to the point where the radome is the main source of clutter, resulting from internal radome reflections [6].

Depolarization

Radome depolarization is a folding of energy from the primary antenna polarization to the other sense. For instance, assume that the antenna is receiving a right-hand circularly polarized (RHCP) signal. After the received signal propagates through the radome wall, part of the energy is converted into a left-hand circularly polarized (LHCP) signal. This phenomenon occurs as a result of (1) radome wall curvature and (2) the difference in complex transmission coefficient between orthogonal polarized vectors.

Depolarization can be a big problem, particularly with satellite communications (SATCOM) ground terminals that utilize frequency reuse such as that shown in Figure 1.3. Here two independent signals are transmitted or received within the same frequency channel, but in opposite polarization senses.

Insertion Loss as a Result of the Presence of a Radome

Lastly, radome insertion loss is a reduction in signal strength as the electromagnetic wave propagates through the dielectric radome wall. Part of the loss occurs as a

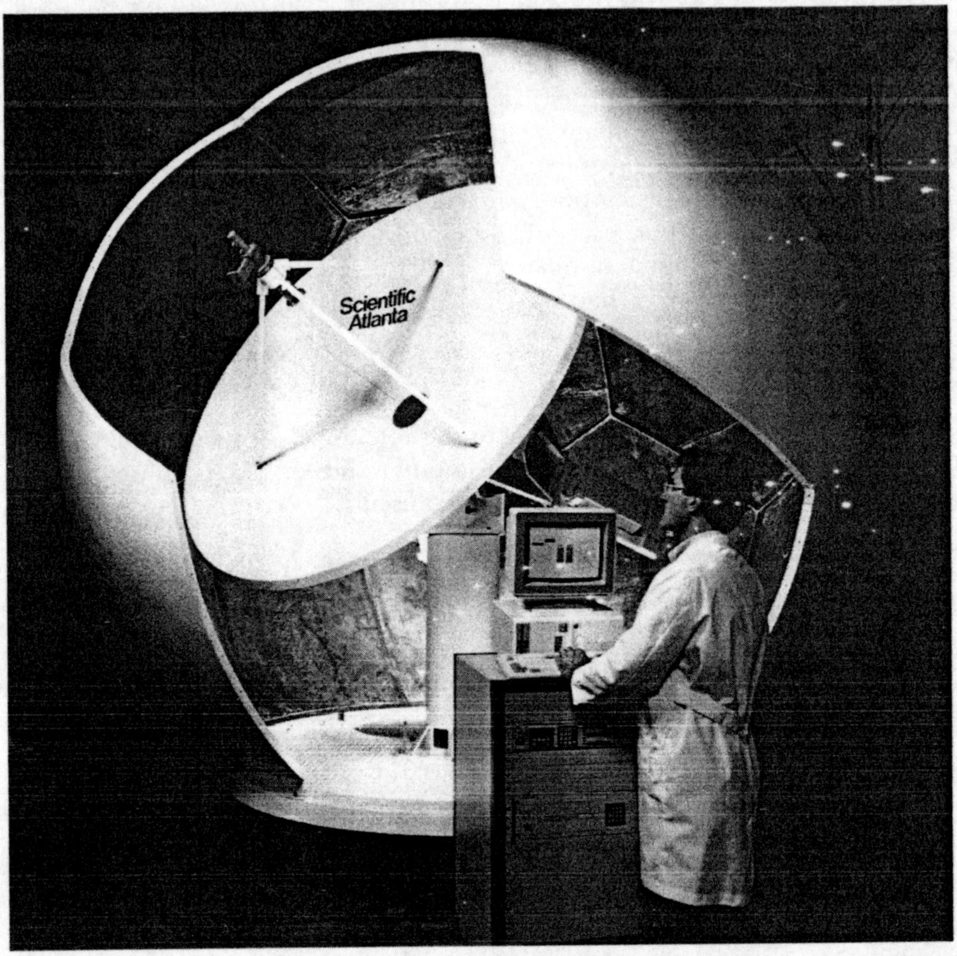

Figure 1.3 Dielectric segmented radome. (Photo courtesy of Scientific-Atlanta, Inc.)

reflection at the air/dielectric interface. The remainder of the loss occurs from dissipation within the dielectric layers. The so-called material loss tangent is a measure of these dissipative losses.

1.4 SIGNIFICANT PARAMETERS IN RADOME ANALYSIS

The particular application decides which radome-induced effects are of concern. Table 1.1 gives an overview of the most common radome applications and the parameters important to the design engineer.

Early methods of radome evaluation used to determine these parameters were cumbersome and approximate, relying primarily on the use of the nomograph [7].

With the arrival of the computer, modern mathematical approaches began to appear, featuring better computational accuracy [8].

For high-accuracy radome performance computations, the efficiency of the calculation determines its limitation as a design tool. Therefore, the combination of accuracy and efficiency must be a goal in any radome/antenna systems analysis approach.

In addition, the radome design community faces increased computer run times for analysis of new millimeter-wave (very-short-wavelength) radar systems. The increased run time is due to the fact that sample spacings are typically a half wavelength over the antenna aperture. This additional run time adds impetus

Table 1.1
Significant Radome Performance Parameters

Application	*Radome Loss*	*Antenna Sidelobe Levels*	*Depolarization*	*Boresight Error*	*Boresight Error Slope*
Ground/Sea					
Weather radars	•	•	•		
Aircraft surveillance radars	•	•	•	•	
Marine radars	•	•	•		
Gun fire control radars	•	•			•
Terrestial (point-to-point) communications	•				
Satellite communications	•	•	•		
Satellite navigations	•	•			
Airborne					
Weather radars	•	•	•		
Radar Altimeters	•				
Radar beacons	•	•	•		
Fire control radars	•	•	•	•	
Aircraft early warning (AWACs)	•	•			•
Satellite communications	•	•	•		
ECM systems	•	•	•		
Missiles/Munitions					
Radar guidance	•	•	•	•	•
Passive antiradiation homing (ARH)	•	•		•	•
Passive radiometric	•	•		•	

to the search for more efficient methods of performing millimeter-wave radome analysis.

Fortunately, computational approaches that would have required a mainframe a few years ago can now utilize the low-cost, high-performance personal computer (PC). The goal of this book is to provide you with working tools amenable to PC modeling so that you can quantify the radome-induced parameters of interest in a particular application.

References

[1] Tice, T. E., "Techniques for Airborne Radome Design," AFATL-TR-66-391, Air Force Avionics Laboratory, Wright Patterson AFB, Ohio, December 1966.

[2] Baxter, J. P., *Scientists Against Time,* Boston: Little Brown and Company, 1952.

[3] Eggleston, W., *Scientists at War,* Toronto: Hunter Rose Company, 1950.

[4] Johnson, R. C., *Seeker Antennas,* Chap. 38 in *Antenna Engineering Handbook,* 3rd ed., New York: McGraw Hill Publishing Company, 1995.

[5] Yueh, W. R., "Adaptive Estimation Scheme of Radome Error Calibration," *Proc. 22nd IEEE Conf. on Decision and Control,* Vol. 8, No. 5, 1983, pp. 666–669.

[6] Rulf, B., "Problems of Radome Design for Modern Airborne Radar," *Microwave J.,* Vol. 28, No. 5, May 1995, pp. 265–271.

[7] Kaplun, V. A., "Nomograms for Determining the Parameters of Plane Dielectric Layers of Various Structure with Optimum Radio Characteristics," *Radiotechnika I Electronika* (Russian), Part 2, Vol. 20, No. 9, 1965, pp. 81–88.

[8] Bagby, J., "Desktop Computer Aided Design of Aircraft Radomes," *IEEE MIDCON '88 Conf. Record,* North Hollywood, CA: Western Periodicals Company, 1988, pp. 258–261.

Chapter 2

Basic Principles and Conventions

This chapter provides a reference and review of topics used in the modeling of radome-enclosed antennas. The following topics are covered in this chapter:

- Vector mathematics;
- Electromagnetic theory;
- Matrices;
- Coordinate systems and antenna pointing gimbal relationships.

These topics are not treated rigorously, but rather in a top-level summary. If you are well-versed in these subjects, you may wish to skip most of this chapter. However, you should review Section 2.4 because it defines reference conventions used in later mathematical modeling.

2.1 VECTOR MATHEMATICS

A Cartesian coordinate system is used where a vector $\overline{A}$ is represented as the sum of three vectors parallel to Cartesian coordinate axes, with the components of $\overline{A}$ given by a_x, a_y, a_z. For this purpose, we associate with that coordinate system three mutually perpendicular unit vectors $\overline{x}$, $\overline{y}$, $\overline{z}$, which have the positive directions of the three coordinate axes. Accordingly, the vector can $\overline{A}$ be represented as

$$\overline{A} = a_x\overline{x} + a_y\overline{y} + a_z\overline{z} \tag{2.1}$$

Figure 2.1 illustrates the three vectors $\overline{x}$, $\overline{y}$, $\overline{z}$ when the origin is chosen as their common initial point. In like manner, another vector $\overline{B}$ may be defined as

$$\overline{B} = b_x\overline{x} + b_y\overline{y} + b_z\overline{z} \tag{2.2}$$

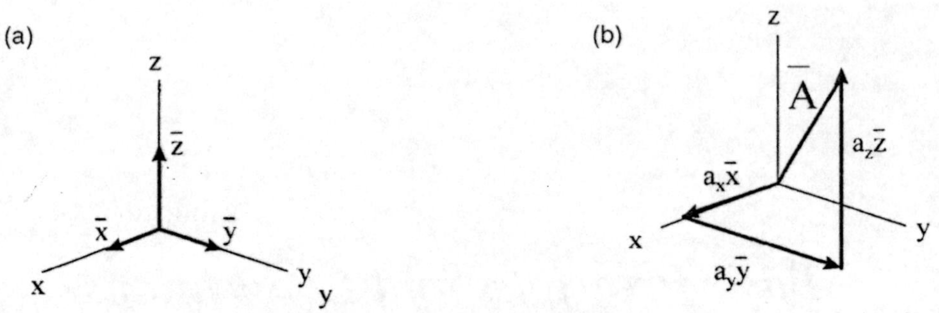

Figure 2.1 (a) The unit vectors $\overline{x}$, $\overline{y}$, $\overline{z}$ and (b) the representation of vector $\overline{A}$.

The dot and scalar cross-products of vectors $\overline{A}$ and $\overline{B}$ are both scalar quantities expressed in terms of the angle ψ between the two vectors [1].

$$\overline{A} \cdot \overline{B} = AB \cos \psi \tag{2.3}$$

$$\left| \overline{A} \times \overline{B} \right| = AB \sin \psi \tag{2.4}$$

where the magnitude of these two vectors are expressed by the following two formulas:

$$A = \sqrt{a_x^2 + a_y^2 + a_z^2} \tag{2.5}$$

$$B = \sqrt{b_x^2 + b_y^2 + b_z^2} \tag{2.6}$$

However, the vector cross-product of vectors $\overline{A}$ and $\overline{B}$ results in another vector, as illustrated in Figure 2.2:

$$\overline{C} = \overline{A} \times \overline{B} \tag{2.7}$$

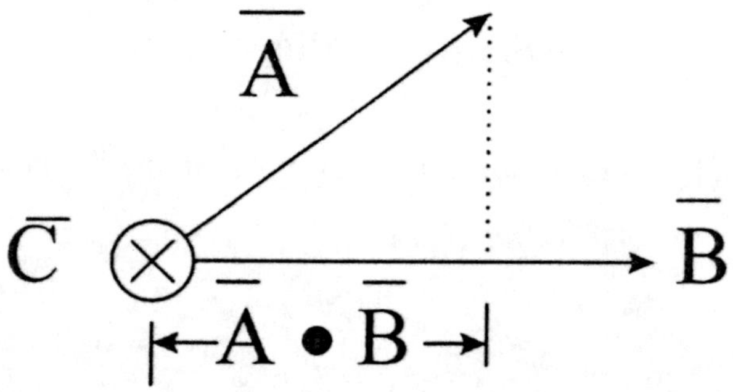

Figure 2.2 Dot and cross-products of vectors $\overline{A}$ and $\overline{B}$.

This vector is orthogonal to the plane in which $\overline{A}$ and $\overline{B}$ lie, and its sense is determined by the right-hand rule. In matrix form, it is given by [2]:

$$\overline{C} = \begin{bmatrix} \overline{x} & \overline{y} & \overline{z} \\ a_x & a_y & a_z \\ b_x & b_y & b_z \end{bmatrix} \tag{2.8}$$

or, expanding the matrix,

$$\overline{C} = (a_y b_z - b_y a_z)\overline{x} + (b_x a_z - a_x b_z)\overline{y} + (a_x b_y - b_x a_y)\overline{z} \tag{2.9}$$

Other vector manipulations that are worth mentioning include the following:

- The *distributive law* for scalar multiplication [3]:

$$\overline{A} \cdot (\overline{B} + \overline{C}) = \overline{A} \cdot \overline{B} + \overline{A} \cdot \overline{C} \tag{2.10}$$

- The *distributive law* for vector multiplication:

$$\overline{A} \times (\overline{B} + \overline{C}) = \overline{A} \times \overline{B} + \overline{A} \times \overline{C} \tag{2.11}$$

- The *scalar and vector triple products*, respectively:

$$\overline{A} \cdot (\overline{B} \times \overline{C}) = (\overline{A} \times \overline{B}) \cdot \overline{C} = \overline{C} \cdot (\overline{A} \times \overline{B}) + \overline{B} \cdot (\overline{C} \times \overline{A}) \tag{2.12}$$

$$\overline{A} \times (\overline{B} \times \overline{C}) = (\overline{A} \cdot \overline{C})\overline{B} - (\overline{A} \cdot \overline{B})\overline{C} \tag{2.13}$$

Finally, some relationships important in electromagnetic theory are as follows:

- The *curl* of a vector:

$$\nabla \times \overline{A} = \begin{bmatrix} \overline{x} & \overline{y} & \overline{z} \\ \dfrac{\partial}{\partial x} & \dfrac{\partial}{\partial y} & \dfrac{\partial}{\partial z} \\ a_x & a_y & a_z \end{bmatrix} \tag{2.14}$$

- The *divergence* of a vector:

$$\nabla \cdot \overline{A} = \frac{\partial a_x}{\partial x} + \frac{\partial a_y}{\partial y} + \frac{\partial a_z}{\partial z} \tag{2.15}$$

- The *curl of the curl* of a vector:

$$\nabla \times (\nabla \times \overline{A}) = \nabla(\nabla \cdot \overline{A}) - \nabla^2 \overline{A} \tag{2.16}$$

2.2 ELECTROMAGNETIC THEORY

Development of the wave equation begins with Maxwell's relationships for the electric and magnetic field quantities applicable to a charge and impressed current free region:

$$\nabla \cdot \overline{D} = 0 \tag{2.17}$$

$$\nabla \cdot \overline{B} = 0 \tag{2.18}$$

$$\nabla \times \overline{E} = -\mu \frac{\partial \overline{H}}{\partial t} \tag{2.19}$$

$$\nabla \times \overline{H} = \epsilon \frac{\partial \overline{E}}{\partial t} \tag{2.20}$$

along with the so-called *auxiliary relations*:

$$B = \mu H \tag{2.21}$$

$$D = \epsilon E \tag{2.22}$$

The preceding differential equations are the fundamental Maxwell relationships, written in a rationalized mks system of units. In this system, we have

E = electric intensity (V/m);
D = electric displacement (C/m^2);
H = magnetic intensity (A/m);
B = magnetic induction (Wb/m);
$\mu = \mu_r \mu_0$ = permeability or magnetic inductive capacity of medium (H/m);
μ_r = relative permeability (dimensionless);
$\mu_0 = 4\pi \times 10^{-7} = 1.257 \times 10^{-6}$ H/m;
$\epsilon = \epsilon_r \epsilon_0$ = permittivity or electric inductive capacity of medium (F/m);
ϵ_r = relative permittivity (dielectric constant) (dimensionless);
$\epsilon_0 = 8.854 \times 10^{-12}$ F/m.

Note that the permittivity is a complex number

$$\epsilon = \epsilon' - j\epsilon'' \tag{2.23}$$

where ϵ' is the real part of ϵ in farads per meter and ϵ'' is the imaginary part of ϵ, also in farads per meter.

Assuming a time harmonic ($e^{j\omega t}$) field, a solution to the wave equation developed from Maxwell's equations for a wave propagating in the z direction is needed for

$$\nabla^2 \overline{E} = -k^2 \overline{E} \tag{2.24}$$

The wave number k is complex for waves propagating through typical radome wall materials since the permittivity is complex.

$$jk = j\omega\sqrt{\mu(\epsilon' - j\epsilon'')} = j\omega\sqrt{\mu\epsilon'\left(1 - j\frac{\epsilon''}{\epsilon'}\right)} \tag{2.25}$$

In the preceding equation, it is convenient to introduce the concept of dielectric material loss tangent:

$$\tan\delta = \frac{\epsilon''}{\epsilon'} \tag{2.26}$$

This concept allows the wave number to be broken into real and imaginary parts [5]:

$$\alpha = \omega\sqrt{\frac{\mu\epsilon'}{2}\left[\sqrt{1+(\tan\delta)^2} - 1\right]} \tag{2.27}$$

$$\beta = \omega\sqrt{\frac{\omega\epsilon'}{2}\left[\sqrt{1+(\tan\delta)^2} + 1\right]} \tag{2.28}$$

Assuming propagation in the z direction, the solution to the wave equation (2.24) is given by

$$E = E_o e^{-jkz} = E_o e^{-\alpha z} e^{-j\beta z} \tag{2.29}$$

In this equation, α is known as the attenuation constant and β as the phase constant. An electromagnetic wave is both attenuated and phase shifted as it propagates through a dielectric.

Valid solutions are orthogonal signal sets E_x, H_y or E_y, H_x related by the wave impedance of the medium

$$\eta = \frac{E_x}{H_y} = \frac{E_y}{H_x} = \sqrt{\frac{\mu}{\epsilon'}}\left[\left(1 - \frac{3(\tan\delta)^2}{8}\right) + j\frac{\tan\delta}{2}\right] \tag{2.30}$$

In terms of polarization, note the following:

1. The polarization is said to be *linear* when only E_x or E_y exists, or when both components exist simultaneously, but are in phase.
2. The polarization is said to be *elliptical* when the two components are out of phase with each other.
3. The polarization is said to be *circular* when the two amplitudes are equal and the phase is +90 or −90 degrees

Because of the modeling used in later sections of this book, an electromagnetic wave incident on a dielectric medium must be quantified into two orthogonal components. These components are defined as parallel and perpendicular to the plane formed by a vector in the direction of propagation and a vector normal to the surface of the radome at the incidence point.

The convention used for parallel and perpendicular wave components for a wave propagating from free space and incident on a dielectric medium is illustrated in Figure 2.3.

2.3 MATRICES

Matrix algebra is a powerful mathematical tool in connection with linear equations and transformations [6].

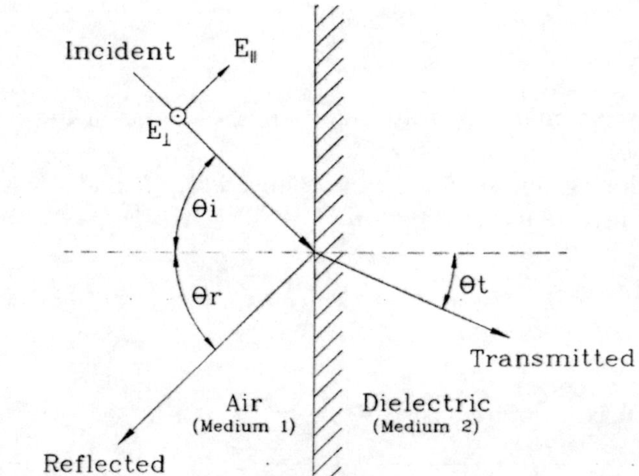

Figure 2.3 Illustration of parallel and perpendicular wave components.

An $m \times n$ matrix is an ordered set of mn quantities, arranged in a rectangular array of m rows and n columns. If $m = n$, the array is a square matrix of order n. A matrix having only one column is called a *column matrix* or *column vector.*

You can prove the following for $m \times n$ matrices $[F]$, $[G]$, $[J]$, and arbitrary constants u and v:

$$[F]+[G]=[G]+[F] \tag{2.31}$$

$$u(v[F])=(uv)[F] \tag{2.32}$$

$$(u+v)[F]=u[F]+v[F] \tag{2.33}$$

$$[F]+([G]+[J])=([F]+[G])+[J] \tag{2.34}$$

$$u([F]+[G])=u[F]+u[G] \tag{2.35}$$

If we have an $m \times n$ matrix $[G]$ and an $n \times p$ matrix $[J]$, the product of $[G]$ and $[J]$ is an $m \times p$ matrix given by

$$[F] = [G][J] \tag{2.36}$$

We can only multiply two matrices if the number of columns of the first factor is equal to the number of rows in the second.

The set of relations given in (2.36) can be solved for $[J]$ via Cramer's rule [7] in order to obtain an inverse transformation

$$[G]^{-1}[G][J] = [J]=[G]^{-1}[F] \tag{2.37}$$

where the identity matrix is defined as

$$[I]=[G]^{-1}[G] \tag{2.38}$$

We now demonstrate the inverse transformation. Consider $[G]$ represented by:

$$[G]=\begin{bmatrix} g_{11} & g_{12} & g_{13} \\ g_{21} & g_{22} & g_{23} \\ g_{31} & g_{32} & g_{33} \end{bmatrix} \tag{2.39}$$

The first step in the process is to form the transpose of $[G]$:

$$[G]^T=\begin{bmatrix} g_{11} & g_{21} & g_{31} \\ g_{12} & g_{22} & g_{32} \\ g_{13} & g_{23} & g_{33} \end{bmatrix} \tag{2.40}$$

To get the inverse of $[G]$, we now replace the elements of the transpose matrix by their corresponding cofactors and divide by the determinant of $[G]$ [7]:

$$[G]^{-1}=\frac{\begin{bmatrix} g_{22}g_{33}-g_{23}g_{32} & g_{13}g_{32}-g_{12}g_{33} & g_{12}g_{23}-g_{13}g_{22} \\ g_{23}g_{31}-g_{21}g_{33} & g_{11}g_{33}-g_{13}g_{31} & g_{13}g_{21}-g_{11}g_{23} \\ g_{21}g_{32}-g_{22}g_{31} & g_{12}g_{31}-g_{11}g_{32} & g_{11}g_{22}-g_{12}g_{21} \end{bmatrix}}{|G|} \tag{2.41}$$

2.4 COORDINATE SYSTEM AND GIMBAL RELATIONSHIPS

The mathematics applied to actual missile or aircraft applications requires that we establish a coordinate system in which to solve the problem. It is necessary to transform the location of an arbitrary point on the antenna face into radome coordinates after all gimbal angular rotations. A derivation of the desired transformations must begin by definition of a standard spherical coordinate system that relates θ, ϕ gimbal rotations into the standard two-axis antenna gimbal AZ, EL rotations that are illustrated in Figure 2.4 for two popular types of aircraft or missile antenna gimbals.

We define these gimbal rotation parameters as follows:

- *Elevation angle (EL):* The angle between the antenna centerline and its projection onto the *z-x* plane of the local coordinate system, often termed the elevation component of the look angle.
- *Azimuth angle (AZ):* The angle between the projection of the antenna centerline onto the *x-y* plane and the boresight, often termed the azimuth component of the look angle.

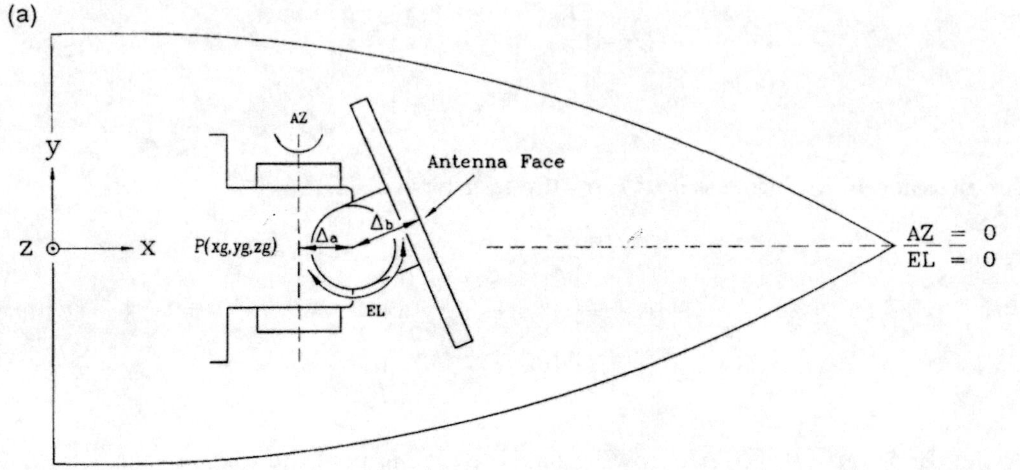

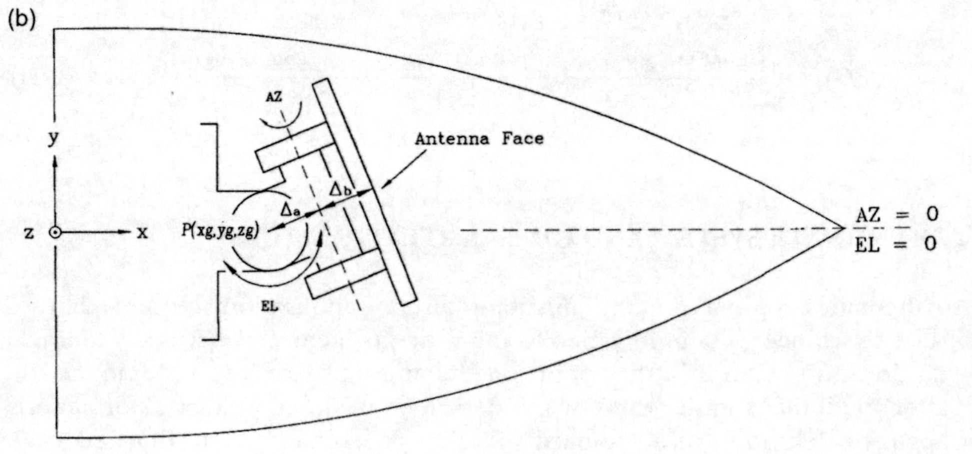

Figure 2.4 (a) Elevation over azimuth (EL/AZ) gimbal and (b) azimuth over elevation (AZ/EL) gimbal.

In missile or aircraft applications, the reference Cartesian coordinate system used throughout this book is generally placed at the geometric center of the radome base in the *y-z* plane, as illustrated in the previous figures. The gimbal location $P_g(x_g, 0, 0)$ within the radome represents the center of gimbal rotations. Note that Δa and Δb are independent azimuth and elevation rotation offsets from $P_g(x_g, 0, 0)$.

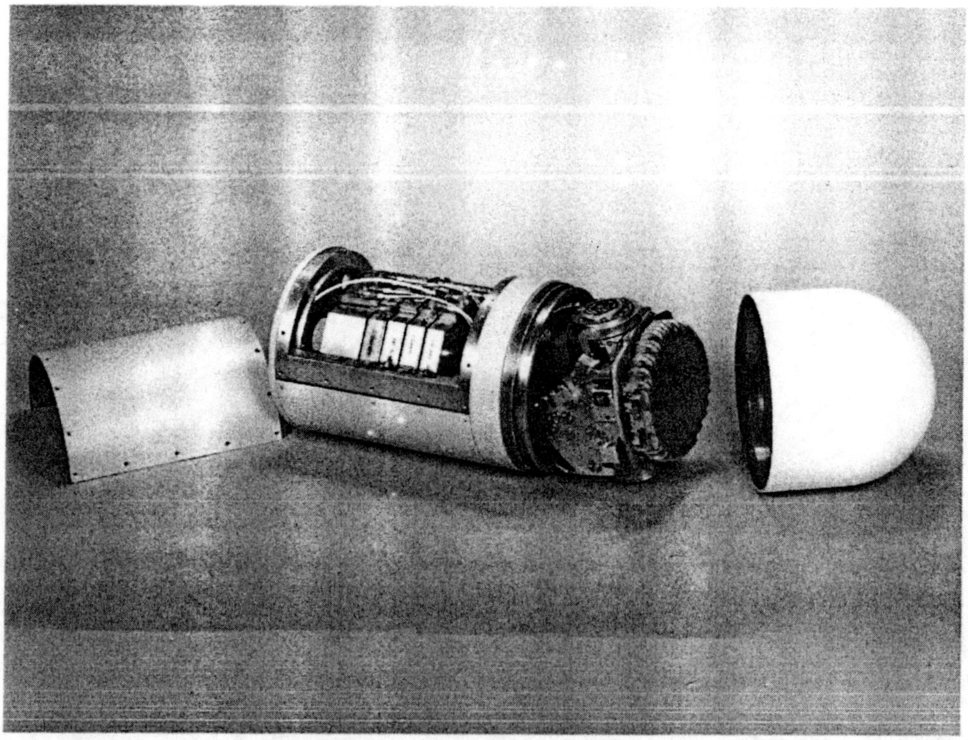

Figure 2.5 Azimuth/elevation gimbaled seeker antenna. (Photograph courtesy of Lockheed Martin Electronics and Missile Systems Company.)

Figure 2.5 is a photograph of a waveguide-type missile-seeker antenna, mounted on an AZ/EL antenna gimbal.

The mathematical development can be simplified by sequentially performing the following steps:

1. With a Cartesian coordinate system placed at the antenna center, compute the transformation that results from angular rotations.
2. Transform results to a Cartesian coordinate system placed at the center of gimbal rotation, taking into account the rotational offsets.
3. Linearly transform the results to true radome coordinates.

In the following development, we will apply these steps.

With a coordinate system placed at the antenna center as illustrated in Figure 2.6, the transformation of any point on the antenna aperture $P_a(x', y', z')$ due to rotations is:

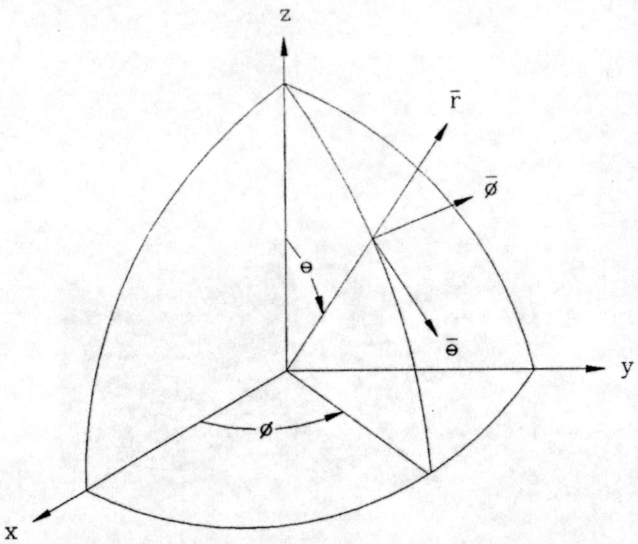

Figure 2.6 Reference spherical coordinate system.

$$\begin{bmatrix} x \\ y \\ z \end{bmatrix} = \begin{bmatrix} \sin\theta\cos\phi & -\sin\theta & -\cos\theta\cos\phi \\ \sin\theta\sin\phi & \cos\phi & -\cos\theta\sin\phi \\ \cos\theta & 0 & \sin\theta \end{bmatrix} \begin{bmatrix} x' \\ y' \\ z' \end{bmatrix} \tag{2.42}$$

Let $\psi = 90$ degrees $- \theta$ then (2.42) becomes

$$\begin{bmatrix} x \\ y \\ z \end{bmatrix} = \begin{bmatrix} \cos\psi\cos\phi & -\sin\phi & -\sin\psi\cos\phi \\ \cos\psi\sin\phi & \cos\phi & -\sin\psi\sin\phi \\ \sin\psi & 0 & \cos\psi \end{bmatrix} \begin{bmatrix} x' \\ y' \\ z' \end{bmatrix} \tag{2.43}$$

The coordinate system space is illustrated in Figure 2.7 for the elevation over azimuth (EL/AZ) antenna gimbal case, and in Figure 2.8 for the azimuth over elevation (AZ/EL) antenna gimbal case.

For the EL/AZ antenna gimbal case, we can now incorporate translations due to the offsets from the gimbal point. We do this by first applying the elevation gimbal rotation with $x' = \Delta_b$, $\psi = 0$ degrees, and $\phi = \text{EL}$. Next, the azimuth offset, Δa, and then the azimuth gimbal rotation is applied by letting $\phi = 0$ degrees and $\psi = \text{AZ}$. We then obtain for the transformation (Rich Matyskiela, AEL Industries, Lansdale, PA, private correspondence, 1994):

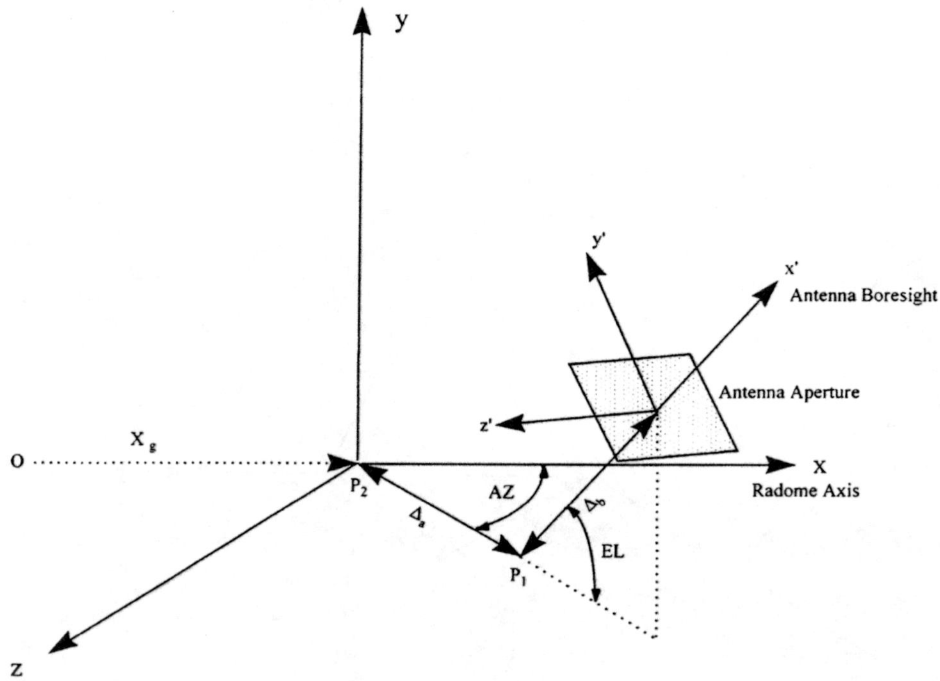

Figure 2.7 Elevation over azimuth (EL/AZ) coordinate system.

$$\begin{bmatrix} x \\ y \\ z \end{bmatrix} = \begin{bmatrix} \cos AZ & 0 & -\sin AZ \\ 0 & 1 & 0 \\ \sin AZ & 0 & \cos AZ \end{bmatrix} \begin{bmatrix} \cos EL & -\sin EL & 0 \\ \sin EL & \cos EL & 0 \\ 0 & 0 & 1 \end{bmatrix} \begin{bmatrix} \Delta b \\ y' \\ z' \end{bmatrix} + \begin{bmatrix} \Delta a \\ 0 \\ 0 \end{bmatrix} \qquad (2.44)$$

Doing the indicated matrix multiplications, and adding a translation for the initial gimbal location $P_g(x_g, 0, 0)$, we obtain the final result:

$$\begin{bmatrix} x \\ y \\ z \end{bmatrix} = \begin{bmatrix} \cos AZ \cos EL & -\cos AZ \sin EL & -\sin AZ \\ \sin EL & \cos EL & 0 \\ \sin AZ \cos EL & -\sin AZ \sin EL & \cos AZ \end{bmatrix} \begin{bmatrix} \Delta b \\ y' \\ z' \end{bmatrix} + \begin{bmatrix} \Delta a \cos AZ \\ 0 \\ \Delta a \sin AZ \end{bmatrix} + \begin{bmatrix} x_g \\ 0 \\ 0 \end{bmatrix} \qquad (2.45)$$

Similarly, for an AZ/EL gimbal, the following matrix solution includes the same translations for the rotational offsets and the initial gimbal location:

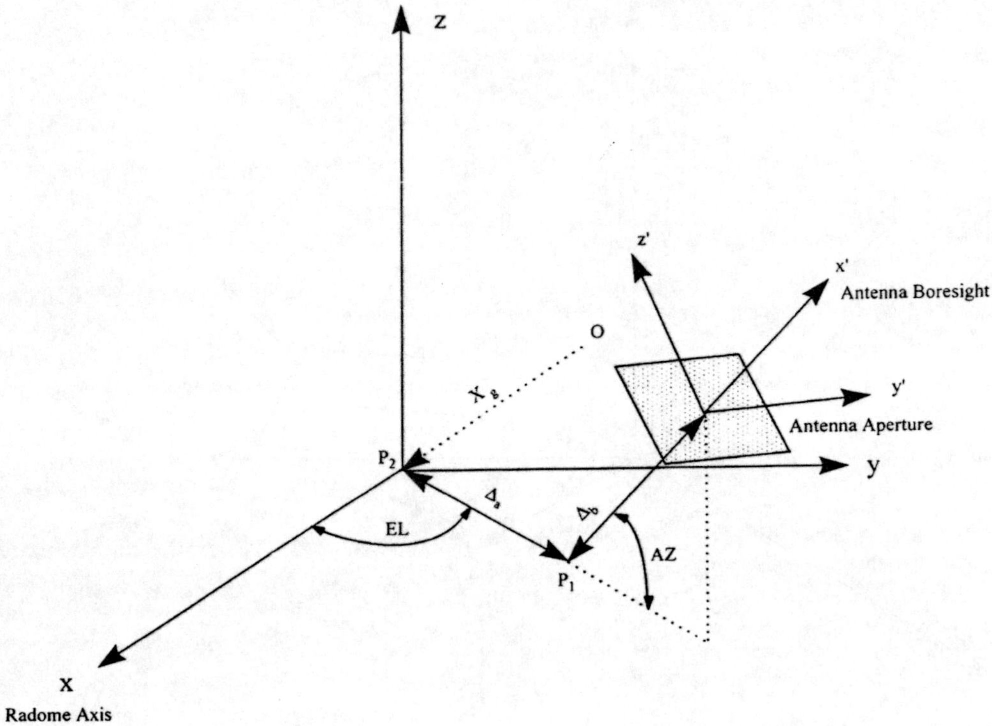

Figure 2.8 Azimuth over elevation (AZ/EL) coordinate system.

$$\begin{bmatrix} x \\ y \\ z \end{bmatrix} = \begin{bmatrix} \cos \text{EL} \cos \text{AZ} & -\sin \text{EL} & -\cos \text{EL} \sin \text{AZ} \\ \sin \text{EL} \cos \text{AZ} & \cos \text{EL} & -\sin \text{EL} \sin \text{AZ} \\ \sin \text{AZ} & 0 & \cos \text{AZ} \end{bmatrix} \begin{bmatrix} \Delta b \\ y' \\ z' \end{bmatrix} + \begin{bmatrix} \Delta a \cos \text{EL} \\ \Delta a \sin \text{EL} \\ 0 \end{bmatrix} + \begin{bmatrix} x_g \\ 0 \\ 0 \end{bmatrix} \tag{2.46}$$

In computer modeling of radome-enclosed gimbaled antenna cases discussed later in this book, we will use these latter two expressions extensively.

References

[1] Sokolnikoff, I. S., and R. M. Redheffer, *Mathematics of Physics and Modern Engineering*, New York: McGraw Hill Book Company, 1958.

[2] Kreyszig, E., *Advanced Engineering Mathematics*, New York: John Wiley & Sons, March 1965.

[3] Reference Data for Radio Engineers, 5th ed., New York: Howard W. Sams & Company, 1974.

[4] Hirsch, H. L., and D. C. Grove, *Practical Simulation of Radar Antennas and Radomes*, Norwood, MA: Artech House, 1987.

[5] Ramo, S., J. R. Whinnery, and T. Van Duzer, *Fields and Waves in Communications Electronics*, 2nd ed., New York: John Wiley & Sons, 1984.
[6] Franklin, P., *Functions of Complex Variables*, Englewood Cliffs, NJ: Prentice Hall, 1958.
[7] Pipes, Louis A., and Lawrence R. Harville, *Applied Mathematics for Engineers and Physicists*, New York: McGraw Hill Book Company, 1970.

Chapter 3

Antenna Fundamentals

In this chapter, we discuss the radiation patterns of an antenna, which follow from the solution of Maxwell's equations. The solution is subject to boundary conditions at both the radiator and at infinity distance. However, most antenna types are too complicated for direct solution. Fortunately, we can make approximations to arrive at a simplified solution.

We will examine a variety of radiators, including aperture and spiral antennas. For these antennas, we can use Huygens's principle to help us analyze the problem. Huygens's principle states that each elemental portion of the wave in the aperture can be considered as a source of waves in space.

This chapter focuses primarily on the antenna and its radiation patterns, but keep in mind that this chapter serves only as background material. The main focus of this book is the impact of the radome on the antenna radiation patterns.

3.1 DIRECTIVITY AND GAIN

Diffraction theory shows that, for an antenna without a radome, the minimum angle within which radiation can be concentrated by an antenna is [1]:

$$\Theta_o \approx \frac{1}{L_a} \tag{3.1}$$

where Θ_o is the half-power beamwidth in radians and L_a is the length of the antenna in wavelengths.

In evaluating an antenna's radiating characteristics, we need only consider the electromagnetic field at a great distance from the antenna. In this far-field

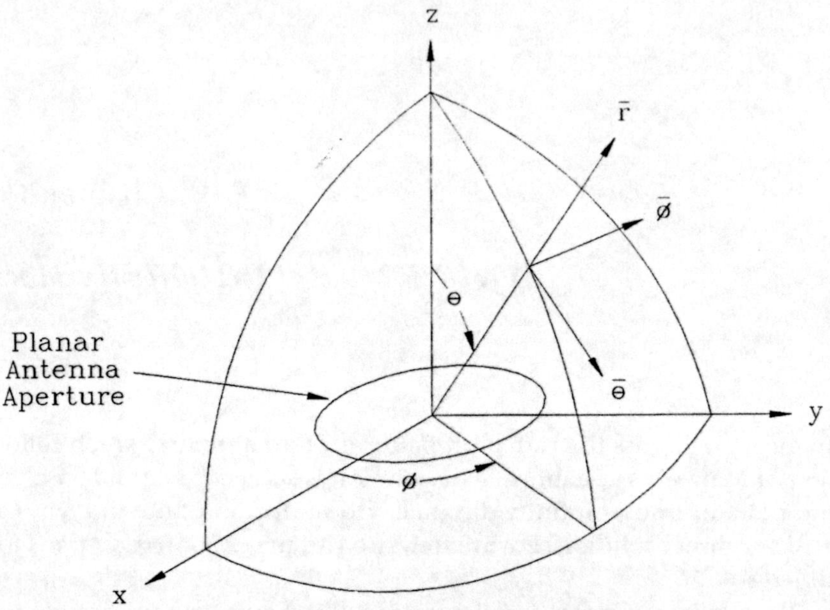

Figure 3.1 Spherical coordinate system centered on antenna aperture.

region, the antenna's power radiation pattern varies only with angles and not with range.

The directive properties of an antenna are expressed as the radiation intensity function, $P(\theta, \phi)$. In this function, θ is the elevation and ϕ is the azimuthal angle in a standard spherical coordinate system centered on the antenna, as illustrated in Figure 3.1. Accordingly, the directive power gain function, G_D, is related to the maximum value of the effective power radiated per unit solid angle at the antenna beam peak, P_{max}, and the total radiated power, P_t, by [2]:

$$G_D = \frac{P_{max}}{P_t / 4\pi} = \frac{4\pi P(0,0)}{\int_0^{2\pi}\int_0^{\pi} P(\theta,\phi) \sin\theta \, d\theta \, d\phi} \tag{3.2}$$

This equation makes two assumptions:

1. We integrate the term in the denominator in all directions of space about the antenna.
2. The maximum radiation is in the direction $\theta = 0$, $\phi = 0$.

For an antenna with 100% radiation efficiency, we relate the antenna directivity to the half-power antenna beamwidths in each principal plane, Θ_o and Φ_o, respectively, by

$$G_D \approx \frac{4\pi}{\Theta_o \Phi_o} \tag{3.3}$$

where both the half-power beamwidths are in radians.

For any physically realizable antenna, the radiation efficiency is less than 100% because of ohmic losses within the antenna structure. We relate the power gain of any antenna to antenna radiation efficiency, η_a, via

$$G = \eta_a G_D \tag{3.4}$$

For aperture antennas, the directive gain and power gain are related to the physical aperture area and effective aperture area, A and A_{eff}, by

$$G_D = \frac{4\pi A}{\lambda^2} \tag{3.5}$$

and

$$G = \frac{4\pi A_{\text{eff}}}{\lambda^2} \tag{3.6}$$

From the last three expressions, we can derive a formula for antenna efficiency, based on the effective area:

$$\eta_a = \frac{A_{\text{eff}}}{A} \tag{3.7}$$

Since the antenna radiation efficiency is equal to or less than unity, we can conclude that the effective antenna area is always equal or less than its physical area. This statement implies that the amount of power a receiving antenna can gather from an incident plane electromagnetic wavefront is generally less than the power impinging on its physical area due to internal losses.

3.2 RADIATION FROM CURRENT ELEMENTS

By its nature, computer modeling of antennas requires antenna aperture currents to be sampled at discrete points as opposed to as a continuous distribution. In the equations to follow, we have assumed a sinusoidally time-varying current of constant frequency, which means that a time factor of $e^{j\omega t}$ is relevant but has not been included.

Consider the far-field radiation patterns for a *z*-directed current element of amplitude I_o and length *d*, as depicted in Figure 3.2. For this infinitesimal dipole, the element radiation patterns at distance *r* from the dipole are [3,4]:

$$H_{\phi} = \frac{I_o d}{2\pi} e^{-jkr}\left(\frac{jk}{r} + \frac{1}{r^2}\right)\sin\theta \tag{3.8}$$

$$E_r = \frac{I_o d}{4\pi} e^{-jkr}\left(\frac{1}{j\omega\epsilon r^3} + \frac{\eta}{r^2}\right)\cos\theta \tag{3.9}$$

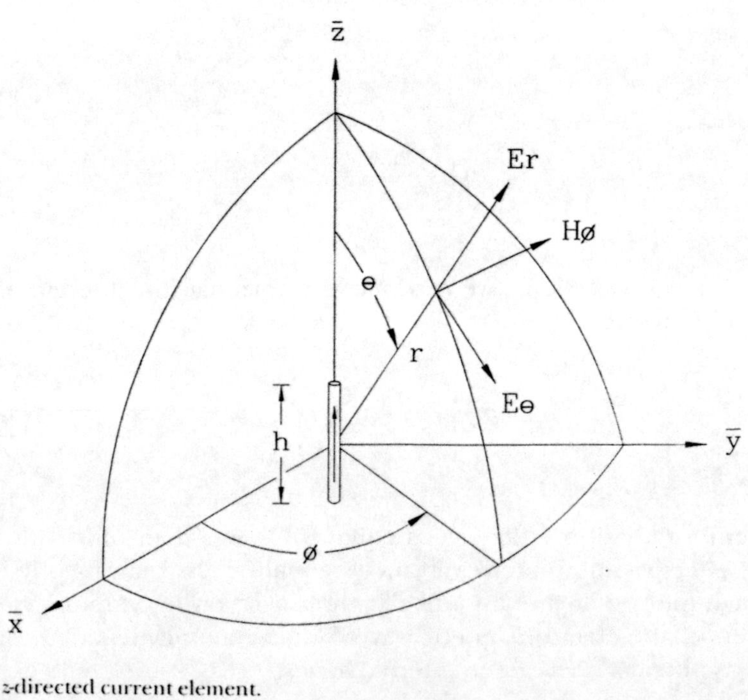

Figure 3.2 A *z*-directed current element.

$$E_\theta = \frac{I_o d}{4\pi} e^{-jkr}\left(\frac{j\omega\mu}{r} + \frac{1}{j\omega\epsilon r^3} + \frac{\eta}{r^2}\right)\sin\theta \tag{3.10}$$

where E in in volts per meter, H is in amperes per meter, and η is the wave impedance in ohms:

$$\eta = \sqrt{\frac{\mu}{\epsilon}} \tag{3.11}$$

For most problems of interest, it is only necessary to know the field components when r is very much greater than a wavelength. For this condition, we may ignore the r^2 and r^3 terms in the preceding expressions. Under the condition where we can do this, we drop the term, E_r. Then, (3.8) and (3.10) reduce to the following:

$$H_\phi \approx \frac{I_o d}{4\pi}\left(\frac{jk}{r}\right)\sin\theta \tag{3.12}$$

$$E_\theta \approx \frac{I_o d}{4\pi}\left(\frac{j\omega\mu}{r}\right)\sin\theta \tag{3.13}$$

From (3.12) and (3.13), the free-space wave impedance is derived as the ratio

$$\eta_o = \frac{E_\theta}{H_\phi} = \sqrt{\frac{\mu_o}{\epsilon_o}} = 120\pi \tag{3.14}$$

At a fixed distance, the variation of the electromagnetic field components only contains a sinusoidal variation with a maximum value in the direction orthogonal to the dipole. When modeling an antenna aperture as an ensemble of dipole radiators, we have to account for this variation of the dipole element pattern.

3.3 ANTENNA ARRAY FACTOR

An antenna element radiates an electric field with two polarization components:

$$E = E_\theta(\theta,\phi)\overline{\theta} + E_\phi(\theta,\phi)\overline{\phi} \tag{3.15}$$

where E_θ and E_ϕ are the two complex electric field components that have a common phase center and $\overline{\theta}$ and $\overline{\phi}$ are unit vectors in the θ and ϕ directions, respectively.

Assume an array of N antenna elements, where N is an integer equal to or greater than two. The total pattern is obtained by adding the electric fields radiated from each, as in the following equation:

$$E = \sum_{i=1}^{N} [E_{\theta i}(\theta,\phi)\overline{\theta} + E_{\phi i}(\theta,\phi)\overline{\phi}] e^{jkr_i} \tag{3.16}$$

We make various approximations in this equation. For instance, we ignore mutual coupling changes in the dipole patterns due to nearby antennas, and only the isolated element patterns are used. If the antennas all have identical element patterns, then we can separate (3.16) into a product [4]:

$$E = [E_\theta(\theta,\phi)\overline{\theta} + E_\phi(\theta,\phi)\overline{\phi}] \sum_{i=1}^{N} I_i e^{jkr_i} \tag{3.17}$$

where E_θ and E_ϕ are the normalized patterns of the electric field from a single element with unity current; I_i is the relative current weight of the ith element and generally includes amplitude and phase terms originating in the feed distribution; and (θ,ϕ) indicates these both have a θ and ϕ dependence.

Equation (3.17) presents a powerful notion because it divides the far-field radiation pattern into an element pattern factor and an array factor. (The element pattern is the left-side expression, and the array factor is the right-side expression.) In the analysis of a radome-enclosed antenna, we are primarily interested in the impact of a radome on the antenna array factor. We will compare the array factor with the radome and the array factor without the radome.

3.4 LINEAR APERTURE DISTRIBUTIONS

Assume that an antenna is transmitting and that it is a linear array of z-directed current elements with a total array length L and centered at L_a, as illustrated in Figure 3.3. If the distribution of radiating current is continuous, the electric and magnetic fields are [5]

$$E_\theta = \left(j\omega\mu \frac{e^{-jkr_o}}{4\pi r_o} \sin\theta \right) \int_0^{L_a} I(z) e^{jkz\cos\theta} dz \tag{3.18}$$

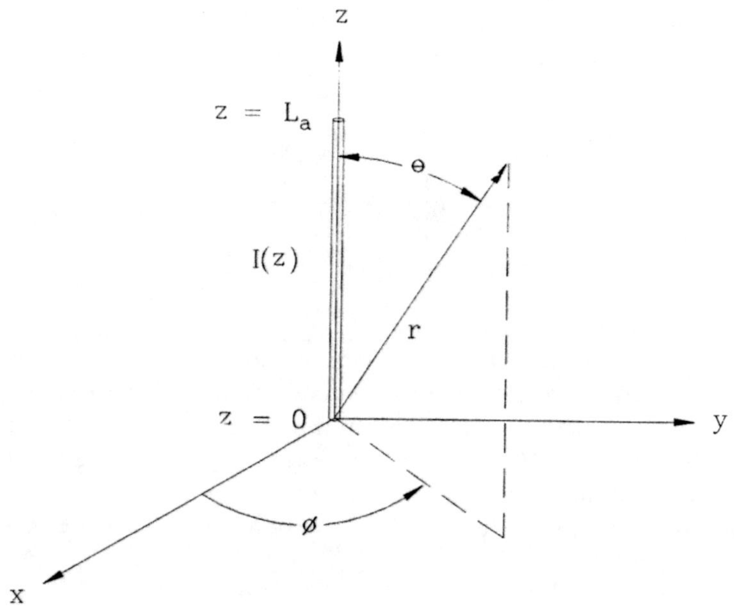

Figure 3.3 Continuous *z*-directed current distribution (line source).

and

$$H_{\phi} = \eta E_{\theta} \tag{3.19}$$

where $I(z)$ is the complex current as a function of z. In (3.18), the integration is over the physical length of the antenna. The array factor is the right side of the expression and is the Fourier transform of the relative current aperture field distribution.

$$E_{\theta}^{\mathrm{AF}} = \int_{0}^{L_a} I(z) e^{j\frac{2\pi z}{\lambda}\cos\theta} dz \tag{3.20}$$

We define the aperture distribution as the current amplitude and phase across the antenna aperture. Controlling the aperture distribution (1) permits sidelobe reduction and (2) allows synthesis of desired antenna beam shapes or control of antenna beam pointing direction (as with a phased-array antenna.)

If the line source has a uniform distribution, $I(z) = 1$, and the far-field radiation pattern is obtained by carrying out the integration indicated in (3.20), then

$$E_\theta^{AF} = L_a \left[\frac{\sin\left(\frac{\pi L_a \cos\theta}{\lambda}\right)}{\frac{\pi L_a \cos\theta}{\lambda}} \right] \tag{3.21}$$

This directivity pattern is significant because it has the highest directivity of all known aperture distributions. However, it has relatively high (−13.2 dB) first sidelobe levels. Figure 3.4 illustrates the computed antenna pattern for a line distribution that is 0.1m long at 10 GHz. In the case of radars, high sidelobes are undesirable because of increased ground clutter reception. High sidelobes are also undesirable in satellite communications because of adjacent satellite interference.

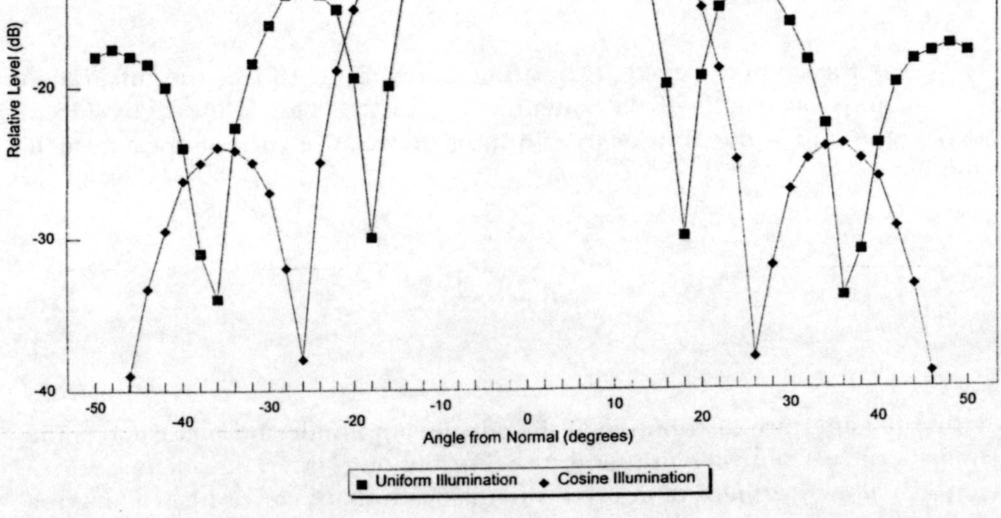

Figure 3.4 Line source antenna patterns (L_a = 0.1m long, f = 10 GHz).

We can reduce antenna sidelobes via tapered antenna aperture current distributions, but only at the expense of increased half-power beamwidth. For instance, consider the following cosine aperture current line distribution:

$$I(z) = \cos\left[\pi\left(\frac{z}{L_a} - \frac{1}{2}\right)\right] \tag{3.22}$$

Substituting this into (3.20) and carrying out the integration, we obtain

$$E_\theta^{\text{AF}} = \frac{L_a}{2}\left[\frac{\sin\left(\frac{\pi L_a}{\lambda}\cos\theta + \frac{\pi}{2}\right)}{\left(\frac{\pi L_a}{\lambda}\cos\theta + \frac{\pi}{2}\right)} + \frac{\sin\left(\frac{\pi L_a}{\lambda}\cos\theta - \frac{\pi}{2}\right)}{\left(\frac{\pi L_a}{\lambda}\cos\theta - \frac{\pi}{2}\right)}\right] \tag{3.23}$$

We show this pattern also in Figure 3.4 for an antenna that is 0.1m long at 10 GHz. It has −23.5-dB sidelobes, but a greater half-power beamwidth than the uniform distribution case.

Other antenna aperture distributions for low sidelobe levels include a cosine raised to a power:

$$I(z) = \cos^q\left(\frac{\pi z}{L_a}\right) \tag{3.24}$$

where q is a power equal to or greater than unity. It need not be an integer.

A parabolic on a pedestal is given by the following formula:

$$I(z) = p - [1 - p]\left[1 - \frac{(z - L_a/2)^2}{(L_a/2)^2}\right] \tag{3.25}$$

where p is the pedestal parameter and is equal to or less than unity. The antenna is centered about $z = L_a/2$.

For all of these distributions, Table 3.1 summarizes the relative gain to the uniform distribution case, the half-power beamwidth, and the first sidelobe levels.

Table 3.1
Line Source Aperture Distribution Pattern Characteristics

Distribution	*Relative Gain (dB)*	*HPBW (deg)*	*1st Null Position (deg)*	*1st Sidelobe (dB)*
Uniform	0.00	$51/(L/\lambda)$	$57.3/(L/\lambda)$	−13.2
$\cos^n$, where:				
$n = 1$	−0.92	$68.75/(L/\lambda)$	$85.94/(L/\lambda)$	−23.0
$n = 2$	−1.76	$83.08/(L/\lambda)$	$114.59/(L/\lambda)$	−32.0
$n = 3$	−2.40	$95.11/(L/\lambda)$	$143.23/(L/\lambda)$	−40.0
$n = 4$	−2.88	$110.58/(L/\lambda)$	$171.89/(L/\lambda)$	−48.0
Parabolic, where:				
$p = 0.8$	−0.03	$52.71/(L/\lambda)$	$60.73/(L/\lambda)$	−15.8
$p = 0.5$	−0.13	$55.58/(L/\lambda)$	$65.32/(L/\lambda)$	−17.1
$p = 0.0$	−0.79	$65.89/(L/\lambda)$	$81.93/(L/\lambda)$	−20.6

A combination of simpler forms can approximate many other distributions. For example, to find the directivity pattern of a cosine raised to a power on a pedestal having the form, use the following:

$$I(z) = p + [1 - p]\cos^q\left[\pi\left(\frac{z}{L_a} - \frac{1}{2}\right)\right] \tag{3.26}$$

The resulting pattern is the superposition of the pattern obtained for a uniform distribution (this corresponds to the pedestal) and the pattern obtained for a cosine raised to the qth power. For instance, for a cosine-squared distribution on a pedestal, the directivity pattern is

$$E_\theta^{AF} = pL_a \frac{\sin\left(\frac{\pi L_a \cos\theta}{\lambda}\right)}{\frac{\pi L_a \cos\theta}{\lambda}} + (1-p)\frac{L_a}{2}\frac{\sin\left(\frac{\pi L_a \cos\theta}{\lambda}\right)}{\frac{\pi L_a \cos\theta}{\lambda}}\frac{\pi^2}{\pi^2 - \left(\frac{\pi L_a \cos\theta}{\lambda}\right)^2} \tag{3.27}$$

In computerized analysis of radome-enclosed antennas, the antenna aperture is generally sampled in discrete increments in accordance with the Nyquist criteria. These criteria are analogous to the criteria for keeping grating lobes out of real space, as in the case of a phased-array antenna. These grating lobes would corrupt antenna pattern data and result in computational errors.

The criterion for determining the maximum sample spacing so that the nearest grating lobe is on the edge of real space corresponds to the condition [6]:

$$\frac{d}{\lambda_o} \leq \frac{1}{1+\sin\theta_o} \tag{3.28}$$

where d/λ_o is the sample-to-sample spacing at the operating frequency and θ_o is the direction of maximum radiation.

Most of the antennas to be addressed in this book have a maximum radiation lobe normal to the antenna aperture surface. This relationship constrains maximum sample spacing to one wavelength at the highest operating frequency.

Assume that there are n samples with d_z being the interval between samples, such that $L_a = Nd_z$. On this basis, the array factor corresponding to (3.20) is

$$E_\theta^{\mathrm{AF}} = \sum_{n=1}^{N} I_n e^{jkz_n \cos\theta} \tag{3.29}$$

where I_n represents the aperture distribution weighting at the nth point and is a complex quantity and z_n is the z coordinate location of the nth point where $z_n = nd_z$.

3.5 TWO-DIMENSIONAL DISTRIBUTIONS

A method similar to that used for the line source can find the far-field radiation directivity pattern for a two-dimensional aperture. However, the Fourier transform is taken over two dimensions instead of one dimension. For instance, if the aperture distribution is over the y-z plane, then the array factor becomes

$$E_\theta^{\mathrm{AF}} = \iint I(y,z) e^{jk\sin\theta(x\cos\phi + y\sin\phi)} dy\, dz \tag{3.30}$$

This integral can also treat circular apertures by defining the appropriate illumination function, $I(y, z)$, to be zero outside of the circular antenna diameter. Table 3.2 shows the far-field antenna beamwidth, first sidelobe level, and gain compared with a uniformly illuminated circular aperture. The aperture has diameter D_a with a distribution of the following form:

Table 3.2
Circular Aperture Distribution Pattern Characteristics

n	*Relative Gain (dB)*	*HPBW (deg)*	*1st Null Position (deg)*	*1st Sidelobe Level (dB)*
0	0.00	$58.44/(D/\lambda)$	$\sin^{-1}[1.22/(D/\lambda)]$	−17.6
1	−1.25	$72.77/(D/\lambda)$	$\sin^{-1}[1.63/(D/\lambda)]$	−24.6
2	−2.52	$84.22/(D/\lambda)$	$\sin^{-1}[2.03/(D/\lambda)]$	−30.6

$$I(y,z)=\left[1-\frac{y^2+z^2}{(D_a/2)^2}\right]^q \tag{3.31}$$

A planar aperture two-dimensional transform separates into one-dimensional transforms when the aperture distribution involves a product where each factor depends on only one coordinate [2], that is,

$$I(y,z)=I(y)I(z) \tag{3.32}$$

In this equation, we define the net radiation pattern to be the product of the principal plane patterns obtained from line source distributions $I(y)$ and $I(z)$, respectively. However, if the aperture distribution is not separable, we must evaluate a two-dimensional integral numerically.

From Maxwell's equations, Appendix A develops a formulation for the far-field radiation patterns of an ensemble of infinitesimal current elements, equally spaced d_y apart in the y plane and d_x apart in the x plane, thus forming a rectangular grid, as illustrated in Figure 3.5. The numerical summation has given the resulting array factor [5]

$$E_\theta^{\mathrm{AF}} = \sum_{m=1}^{M} \sum_{n=1}^{N} I_{mn} e^{jk \sin\theta(md_x \cos\phi + nd_y \sin\phi)} \tag{3.33}$$

The m index pertains to the x coordinate, and the n index to the y coordinate. The weighting coefficient, I_{mn}, usually is a complex quantity. As with the continuous aperture distributions, by proper choice of the coefficients, the boundary of the antenna aperture can be modeled as square, rectangular, elliptical, or circular. For circular, we often assume that we can subdivide the array into four symmetrical quadrants, resulting in excitations that will give sum and difference antenna patterns.

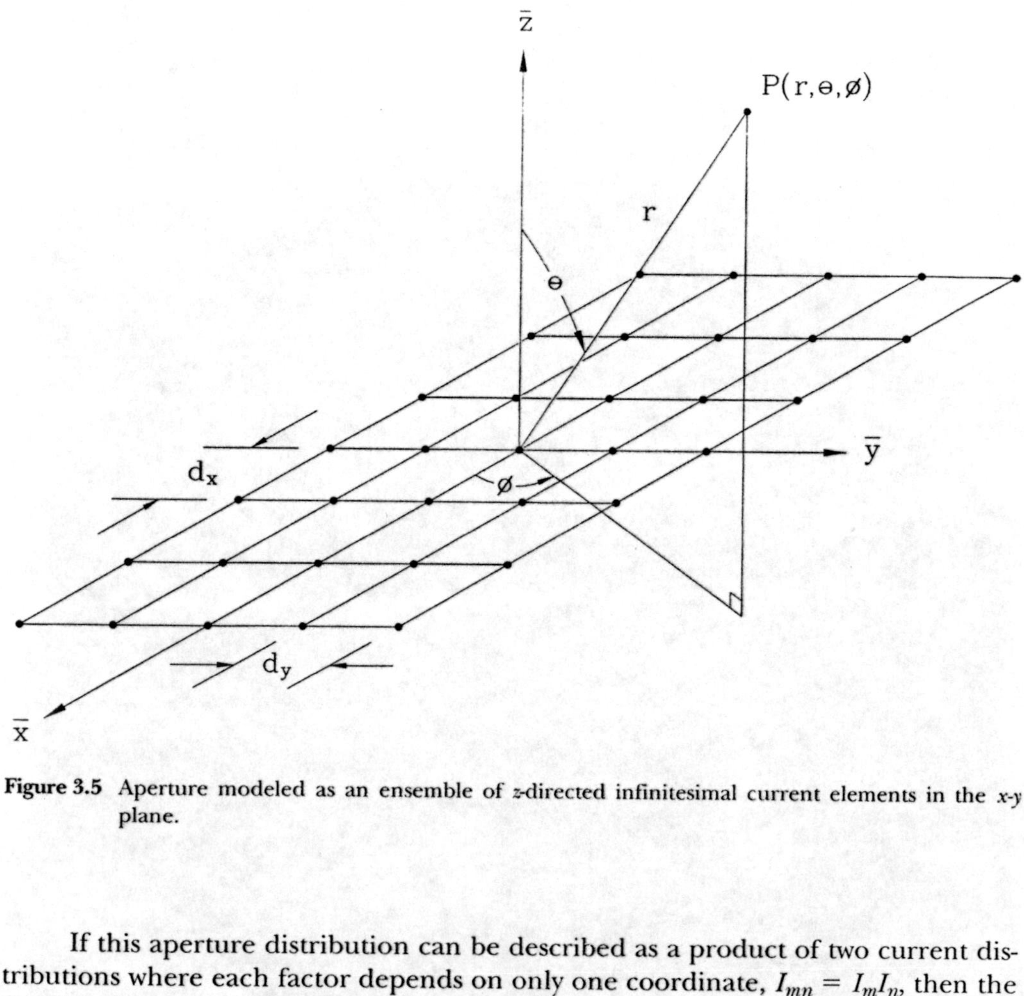

Figure 3.5 Aperture modeled as an ensemble of *z*-directed infinitesimal current elements in the *x-y* plane.

If this aperture distribution can be described as a product of two current distributions where each factor depends on only one coordinate, $I_{mn} = I_m I_n$, then the function is product separable, as indicated in the discussion of (3.33). The net radiation pattern is the product of the principal plane patterns obtained from I_m and I_n, respectively.

$$E_\theta^{AF} = \sum_{m=1}^{M} I_m e^{jkmd_x \sin\theta\cos\phi} \sum_{n=1}^{N} I_n e^{jknd_y \sin\theta\cos\phi} \tag{3.34}$$

where I_m and I_n are complex coefficients that represent the antenna current distributions along the x and y axes. Sometimes these may conveniently correspond to the physical location of radiating elements on a planar antenna aperture. For instance, see the location of slots on the flat-plate waveguide antenna array shown in Figure 3.6.

Figure 3.6 Waveguide flat-plate antenna. (Photograph courtesy of Hughes Missile Systems Division.)

3.6 SPIRAL ANTENNAS

Spiral antennas have frequency-independent radiation patterns and are used in radome-enclosed direction finding (DF) systems on aircraft and on antiradiation homing (ARH) missile guidance systems. An overview of the historical development of these antennas appears in [7–11].

We generally fabricate planar spiral antennas on a printed circuit board. To obtain a unidirectional pattern, a lossless cavity is placed on one side of the spiral. This cavity limits the frequency-independent pattern bandwidth to about an octave. Hamel and Scherer [12] have found that an absorber-lined cavity can extend a spiral operation to a decade bandwidth.

A two-arm (or log-spiral) antenna, shown in Figure 3.7(a), is frequency independent because angles entirely describe the antenna's shape, which make it invariant to a change of scale [11,13,14]. The radius function can be found from the formula

$$\rho = \rho_0 e^{a\Psi} \tag{3.35}$$

where a is the so-called expansion coefficient. We rotate the curve of the equation 180 degrees around the origin to define the second arm of the spiral.

We may analyze the planar spiral structure as a strip radiator with currents flowing along the arms. The inner part of the spirals serves as a nonradiating transmission

(a)

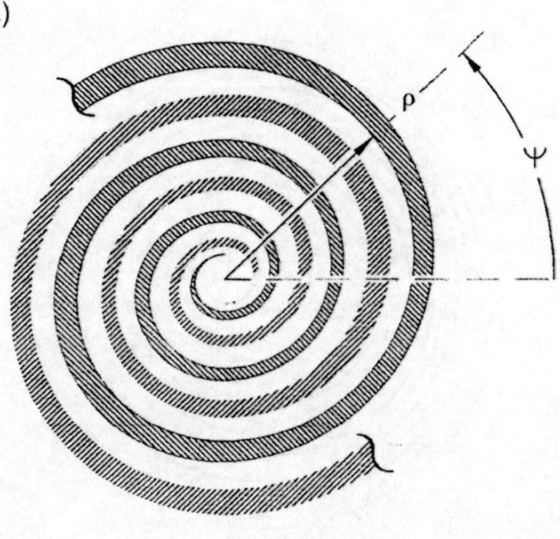

Figure 3.7 Planar spiral antenna geometries: (a) equiangular spiral and (b) Archimedean spiral.

line, carrying the current out to the active region, which is a narrow circumferential ring where antenna radiation takes place. The active regions where radiation takes place correspond to circumferences of $m\lambda$ in diameter, where m is an integer equal to or greater than unity, and is referred to as the *mode* of the radiation.

So-called hand rules determine the sense of the circular polarization. For instance, pointing your thumb away from the spiral, let the fingers roll in the direction of the increasing spiral radius. The hand that must be used determines the polarization sense. For instance, if you must use your right hand in order for your fingers to roll in the direction of increasing spiral radius, then the spiral antenna is of the right-hand circular polarization (RHCP) sense.

The Archimedean spiral antenna has constant spacing between adjacent arms, as illustrated in Figure 3.7(b). It is only frequency independent over a typical decade-bandwidth frequency range [15]. However, to obtain this complete bandwidth, it requires lossy material on the ends of the arms to eliminate end reflections [16].

The Archimedean spiral has a radius function that increases uniformly with the angle Φ according to

$$\rho = \rho_o + a\Psi \tag{3.36}$$

For sufficiently large spiral antennas that support higher order modes, the currents in these modes vary as $e^{jm\psi}$ with Ψ varying from 0 to 2π. All modes for which $m > 1$ have

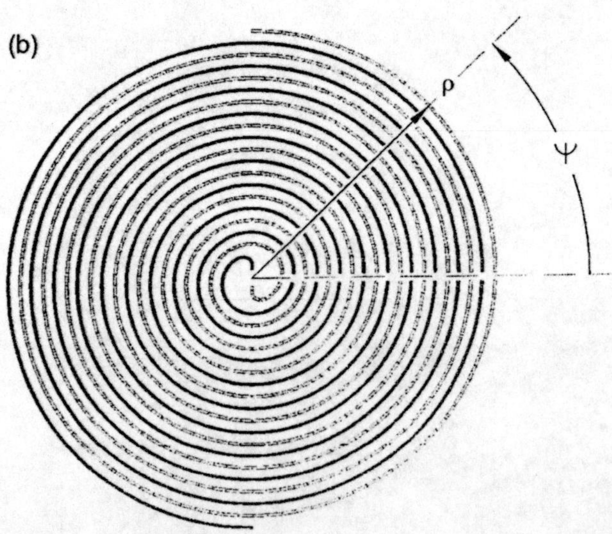

Figure 3.7 (continued)

a null on boresight. For bandwidths less than 3:1 for a two-arm spiral, the 3λ diameter ring cannot exist, so that only the principal $m = 1$ mode exists. For greater electrical bandwidths, control of radiation in the higher order modes can be controlled by spiral design methods; an excellent discussion appears in Corzine and Mosko [17].

With four-arm spirals, it is possible to simultaneously excite and isolate both the mode 1 and mode 2 responses, which appear much like a monopulse antenna sum and difference pattern. We model the spiral by replacing the aperture with two concentric current rings and their proper phase distributions. A phasing technique to obtain these modes is shown in Schuchardt and Kozakoff [18], as illustrated in Figure 3.8. Each of these modes is rotationally symmetric about the spiral axis, and the field is pure circular polarization, with the polarization sense determined by the spiral winding.

For all planar spiral antennas operating in the principal ($m = 1$) mode, the antenna must have an outer circumference of 1.25 wavelengths at the lowest operating frequency. This circumference ensures good antenna patterns and impedance match when the antenna is operating at its lowest frequency.

References

[1] Silver, S., *Microwave Antenna Theory and Design*, New York: McGraw Hill Book Company, 1949.

[2] Jasik, K., *Fundamentals of Antennas*, Chap. 1 in *Antenna Engineering Handbook*, 3rd ed., R. C. Johson, Ed., New York: McGraw Hill Book Company, 1993.

[3] Kraus, J. D., *Antennas*, 2nd ed., New York: McGraw Hill Book Company, 1988.

[4] Milligan, T. A., *Modern Antenna Design*, New York: McGraw Hill Book Company, 1986.

[5] Weeks, W. L., *Antenna Engineering*, New York: McGraw Hill Book Company, 1968.

[6] Mailloux, R. J., *Phased Array Antenna Handbook*, Norwood, MA: Artech House, 1994.

[7] Jordan, E. C., G. A. Deschamps, J. D. Dyson, and P. E. Mayes, "Developments in Broadband Antennas," *IEEE Spectrum*, Vol. 1, April 1964, pp. 58–71.

[8] Wang, J. J. H., and V. K. Tripp, "Design of Multi-Octave Spiral-Mode Microstrip Antennas," *IEEE Trans. Antennas Propagation*, Vol. 39, No. 3, March 1991, pp. 332–335.

[9] Cubley, H. D., and Hayre, H. S., "Radiation Field of Spiral Antennas Employing Multimode Slow Wave Techniques," *IEEE Trans. Antennas Propagation*, Vol. AP-19, January 1971, pp. 126–128.

[10] Dyson, J. D., "The Characteristics and Design of the Conical Spiral Antennas," *IEEE Trans. Antennas Propagation*, Vol. AP-13, July 1965, pp. 488–499.

[11] Deschamps, G. A., and Dyson, J. D., "The Logarithmic Spiral in a Single-Aperture Multi-Mode Antenna System," *IEEE Trans. Antennas Propagation*, Vol. AP-19, January 1971, pp. 90–95.

[12] Du Hamel, R. H., and J. P. Scherer, "Frequency Independent Antennas," Chap. 14 in *Antenna Engineering Handbook*, 2nd ed., R. C. Johnson, Ed., New York: McGraw Hill Book Company, 1993.

[13] Rumsey, V. H., "Frequency Independent Antennas," *IRE National Convention Record*, Part 1, 1957, pp. 114–118.

[14] Dyson, J. D., "The Equiangular Spiral," *IRE Trans. Antennas Propagation*, Vol. AP-7, No. 2, April 1959, pp. 181–187.

[15] Turner, E. M., "Spiral Slot Antenna," U.S. Patent 2,863,145, December 1958.

[16] Rudge, A. W., K. Milne, A. D. Oliver, and P. Knight, *The Handbook of Antenna Design*, 2nd ed., London, Peter Peregrinus Ltd, on behalf of the Institution of Electrical Engineers (UK), 1986.

[17] Corzine, R. G., and J. A. Mosko, *Four Arm Spiral Antennas*, Norwood, MA: Artech House, 1990.

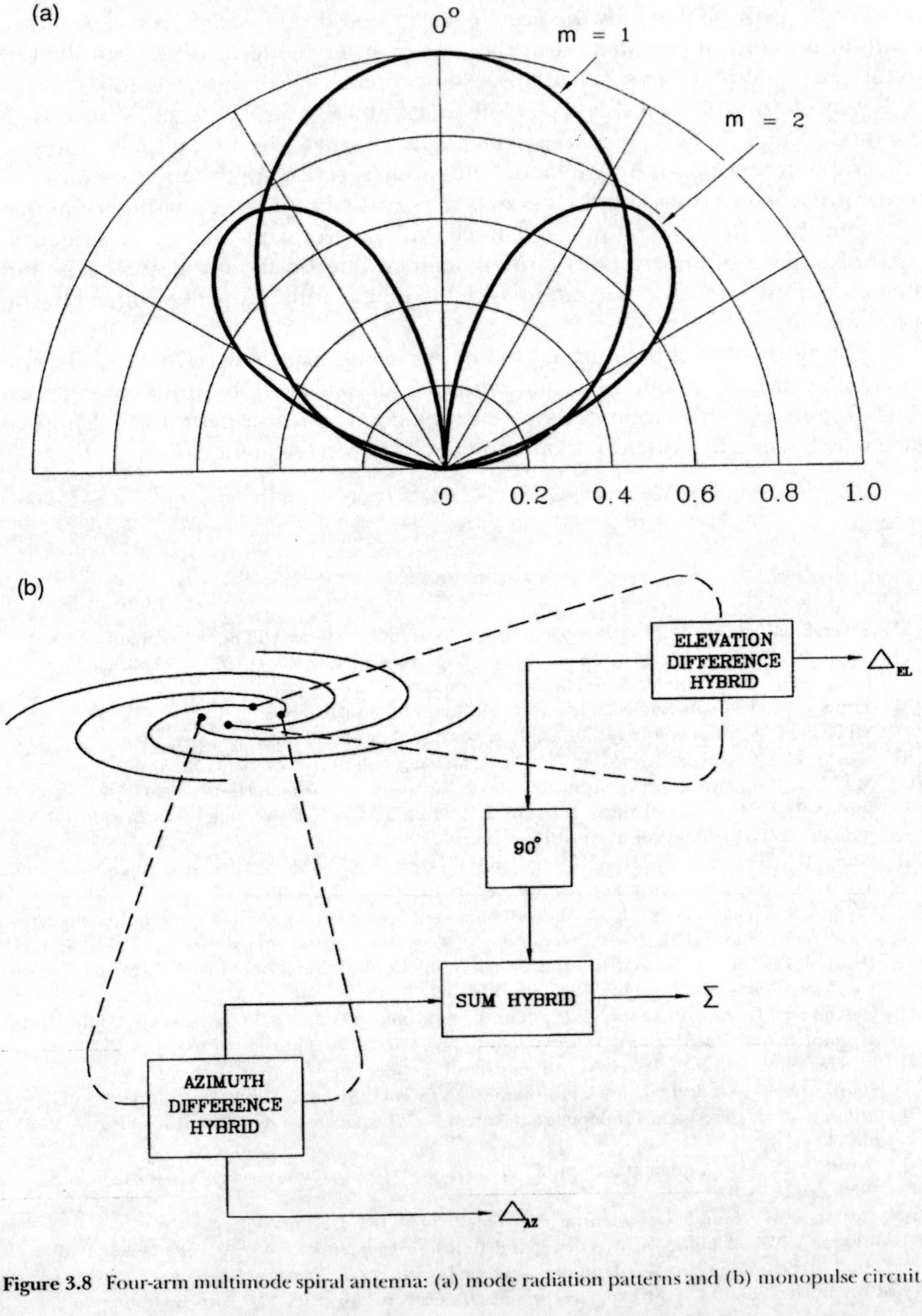

Figure 3.8 Four-arm multimode spiral antenna: (a) mode radiation patterns and (b) monopulse circuit.

[18] Schuchardt, J. M., and D. J. Kozakoff, "Seeker Antennas," Chap. 38 in *Antenna Engineering Handbook,* 3rd ed., R. C. Johnson, Ed., New York: McGraw Hill Book Company, 1993.

Selected Bibliography

Elliot, R. S., *Antenna Theory and Design,* Englewood Cliffs, NJ: Prentice Hall, 1981.

Ramo, S., J. R. Whinnery, and T. Van Duzer, *Fields and Waves in Communications Electronics,* 2nd ed., New York: John Wiley & Sons, 1984.

Chapter 4

Radome Dielectric Materials

In analyzing radome electrical performance, it is important to evaluate the electrical properties of possible radome wall materials at various wavelengths. These properties include (1) relative dielectric constant and (2) loss tangent.

The structural and environmental requirements determine other parameters for a radome design. These other parameters include the following:

- Mechanical properties, such as flexural moduli, strength, and hardness;
- Material density;
- Water absorption;
- Rain erosion resistance;
- The effects of thermal shock.

This chapter focuses on the electrical properties of selected radome wall materials as they influence a radome electrical design. Mechanical and physical data for these materials are readily available in many sources, such as the biannual *Proceedings of the Symposium on Electromagnetic Windows* [1–9]. A survey of the properties of many avionics radome materials was published in a NATO-funded report [10], a survey of ceramics materials in [11], and a review of moldable organic (plastic) materials in a United States Air Force publication [12]. Most recently, the IIT Research Institute and Purdue University reported the preparation of an electromagnetic window material data handbook [13]. This handbook will address visible, infrared (IR), millimeter-wave, and dual-mode materials.

4.1 ORGANIC RADOME DIELECTRIC WALL MATERIALS

Most civilian and military aircraft, ground vehicles, and fixed terrestrial radomes use organic radome wall materials. Radome wall construction is typically of one of the following designs:

- Solid (monolithic), generally made of resins incorporating reinforcements such as chopped glass fibers;
- Sandwich design, consisting of alternating high-density (high relative dielectric constant) and low-density (low relative dielectric constant) materials.

We obtain the resulting relative dielectric constant of the mixture from the following equation:

$$\epsilon_m = \frac{V_R \log \epsilon_R + V_F \log \epsilon_F}{V_R + V_F} \tag{4.1}$$

where

ϵ_m = relative dielectric constant of mixture;
ϵ_R = relative dielectric constant of the resin;
ϵ_F = relative dielectric constant of reinforcement fibers;
V_R = volume of the resin;
V_F = volume of reinforcement fibers

This equation assumes the mixture is uniform and isotropic.

Occasionally, particularly at millimeter-wave bands, monolithic radomes are used without reinforcement. Table 4.1 shows the dielectric properties of several of these materials [14,15].

Sandwich Wall Radomes

Throughout the microwave spectrum, sandwich radomes that include low-density core materials are popular. To fabricate these radomes, the most common method is vacuum bag or autoclave molding.

Table 4.1
Approximate Dielectric Properties of Some Common Plastics

Material	*Relative Permittivity at X-Band*	*Loss Tangent at X-Band*	*Relative Permittivity at K_a-Band*	*Loss Tangent at K_a-Band*
Rexolite	2.54	0.0005	2.60	0.002
Nylon	3.03–3.21	0.014–0.02	3.60	0.02
Teflon	2.10	0.0003	2.08–2.10	0.0006–0.001
Polystyrene	2.55	0.0004	2.54	0.00053–0.001
Plexiglass	2.59	0.015	2.61	0.02
Polyethylene	2.25	0.0004	2.28	0.007

For vacuum bag fabrication, the so-called wet layup was one of the earliest methods for fabrication of multilayer wall radome structures. For an A-sandwich radome, this technique involves laying down plies of the reinforcement fabric that are wetted with a liquid resin. After using a number of plies to build up the desired inner skin thickness, we insert a low-density core layer and repeat the wet-layup process for the outer skin. Subsequently, a plastic bag is placed over the complete layup, thereby sealing the mold. The plastic bag is connected to a vacuum pump that evacuates the air between the bag and the layup. By this fabrication process, the tolerance on skins is typically −1 mil on the buildup of the plies.

Currently more popular in the radome industry is the "pre-preg" fabrication process, which uses pre-impregnated fabric-reinforced laminates. The sheet laminate comes with precise and uniform resin content through the wall. Thus, tolerances can be controlled to a desired thickness more accurately than with the wet-layup approach. Pre-preg material is commercially available in commonly used discrete thicknesses that are multiples of one ply, that is, about 11 mils per ply. Special-order plies are available in nonstandard thicknesses of one-half ply, that is, 5.5 mils per ply.

Table 4.2 shows the approximate dielectric properties of popular fiber reinforcement and resin systems [10,11; Ben MacKenzie, Norton Performance Plastics Corporation, Ravenna, OH, private correspondence, 1996].

Table 4.2
Approximate Dielectric Properties of Selected Fiber and Resin Systems (X-Band)

Material Category	*Reinforcement or Resin Type*	*Relative Permittivity*	*Loss Tangent*
Reinforcement	E-Glass	6.06	0.004
	S-Glass	5.2	0.007
	D-Glass	4	0.005
	Polyethylene (Spectra)	2.25	0.0004
	Kevlar	4.1	0.02
	Quartz	3.8	0.0001
Resin	Polyester	2.95	0.007
	Bismaleimide	3.32	0.004
	Polybutadiene	3.83	0.015
	Epoxy	3.6	0.04
	Polyamide	3.1	0.0055
	Polycyanate	2.86	0.005

The radome designer must be concerned about both the electrical and mechanical properties when selecting a reinforcement. Some of these properties are given in the following list.

- S-glass, a silica-alumina-magnesia composite, is a high-strength glass in which the reinforcement fibers' tensile strength is maintained to 650°C.
- E-glass contains alumina-borosilicate and is low cost. Compared with S-glass, it shows a greater reduction of tensile strength at 650°C.
- D-glass was developed for radome applications and has a lower dielectric constant and loss tangent that the other glass fibers. However, it is lower in strength and higher in cost.
- Spectra consists of ultra-high-strength polyethylene fibers. It is lightweight with high impact resistance, low water pickup, and excellent electrical properties for a low-loss radome.
- Quartz has a high silica content, resulting in a low dielectric constant and a small loss tangent. It is higher in cost, but offers lower loss potential than a polyethylene radome.
- Kevlar is useful for high-strength applications. However, its loss tangent is higher.

When integrating a reinforcement with a resin, the overall electrical properties depend on the ratio of a resin to reinforcement. A mix formula similar to that of (4.1) must determine the resulting electrical properties. Table 4.3 shows the approximate dielectric properties of selected pre-preg laminates available in the industry [10,11; Ben MacKenzie, Norton Performance Plastics Corporation, Ravenna, OH, private correspondence, 1996]. For subsonic aircraft or terrestrial sandwich radomes, the variation of the dielectric properties with operating temperatures is very small [10].

Core Materials

Honeycombs are the ideal basic configuration for lightweight but extremely strong materials, and are the most commonly used A- and C-sandwich core materials. Figures 4.1 and 4.2 illustrate these configurations. Reinforced plastic honeycomb materials are available in a variety of wall resins and reinforcements that operate over a wide temperature range.

Nomex honeycomb has been the leading structural honeycomb core in the aerospace industry. Nomex operates to 350°F, is an aramid honeycomb core material, and has a relative dielectric constant of the order of 1.13 and loss tangent on the order of 0.005.

Korex honeycomb is available in small cell sizes and low densities like the honeycomb of Nomex aramid. We reinforce it with high modulus aramid fibers and a phenolic resin. However, stiffness in shears and compression of the core is about double that of aramid cores. Its electrical properties are the same as Nomex.

Table 4.3
Approximate Dielectric Properties of Selected Laminates (X-Band)

Reinforcement	*Resin*	*Relative Permittivity*	*Loss Tangent*
E-glass	Epoxy	4.4	0.016
	Polyester	4.15	0.015
	Polyimide	4.7	0.014
D-glass	Polycyanate	3.45	0.009
Polyethylene	Epoxy	2.8	0.004
(Spectra)	Polyester	2.52	0.007
	Polycyanate	2.65	0.003
Kevlar	Polyester	3.5	0.05
Quartz	Epoxy	3.12	0.011
	Polyester	3.6	0.012
	Polyimide	3.34	0.005
	Polycyanate	3.23	0.006
	Polybutadiene	3.1	0.003
	Bismaleimide	3.35	0.009

Figure 4.1 Phenolic honeycomb core material.

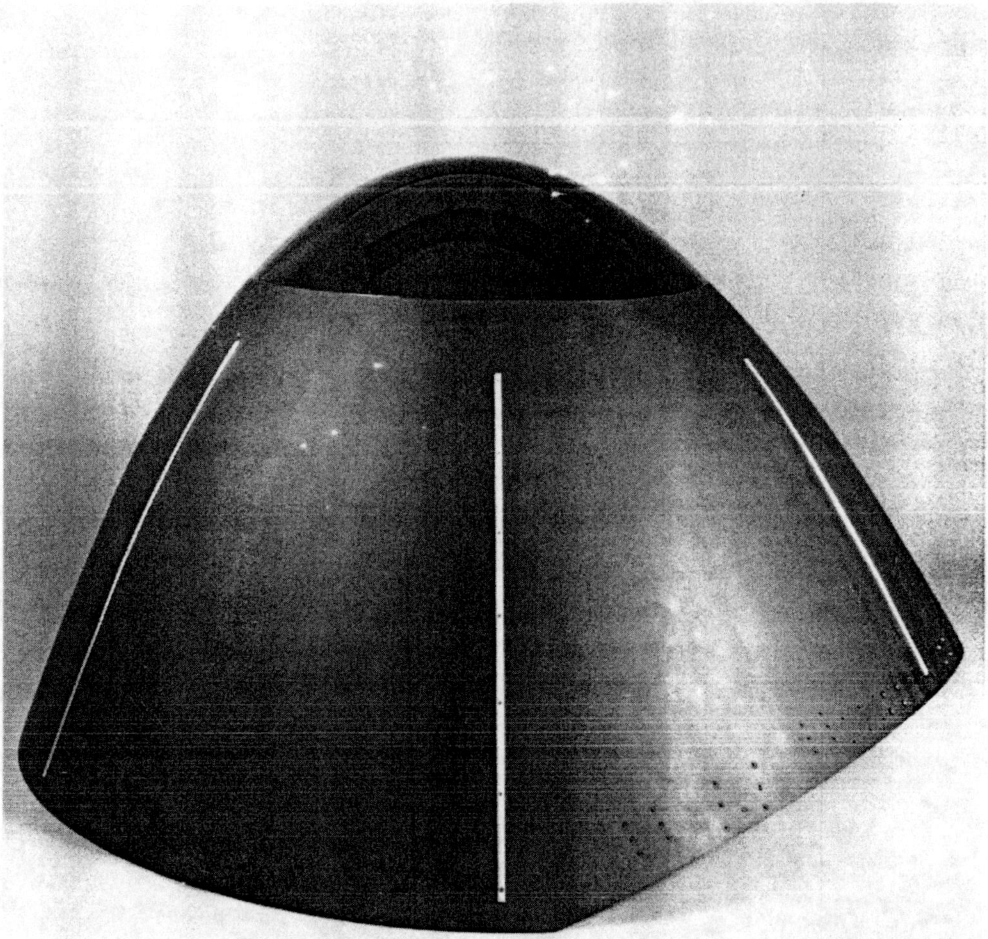

Figure 4.2 A-sandwich aircraft radome using phenolic honeycomb cores. (Photograph courtesy of Norton Performance Plastics Corporation.)

Standard glass-reinforced honeycombs with phenolic resins operate to 350°F. These honeycombs have a relative dielectric constant of the order of 1.35 and loss tangent of 0.001. Anisotropy of honeycomb cores can become a problem at millimeter-wave frequencies.

Many other honeycomb materials are available for higher temperature operation, such as glass reinforcement with a polyimide resin system, which operates to 500°F.

Polyurethane foams have been used as a core material for sandwich radomes.

The flexibility and toughness of the foam is dependent on its density. However, those typically used for radomes are in the range from 0.21 to 0.5 g/cm^3 . This corresponds to relative dielectric constants ranging from about 1.05 to 1.30, and loss tangents ranging from 0.0005 to 0.001, respectively [16]. The dielectric properties of the foam materials are relatively insensitive to frequency well into millimeter-wave regions of the spectrum. Foam materials have been used at over 100 GHz.

4.2 INORGANIC RADOME DIELECTRIC WALL MATERIALS

The mechanical strength of most organic wall materials deteriorates at 250°C, and even the best can only survive for a short time at 500°C [17]. Therefore, high-temperature applications, such as hypersonic missiles, often use nonorganic (ceramic) type radome materials.

Table 4.4 lists representative values for nonorganic radome materials. Most of these materials have suitable electrical properties for high-velocity radome applications, such as missiles. For instance, alumina (aluminum oxide), Pyroceram, and Rayceram (both cordierite) have been widely used for flight vehicles operating at and below about Mach 4. They are hard and have fair rain erosion resistance, but are difficult to grind to shape. Pyroceram has a higher dielectric constant than either Rayceram or alumina, which implies tighter mechanical tolerances in manufacture.

Table 4.4
Dielectric Properties of Selected Ceramic Materials (X-Band)

Materials	*Relative Permittivity*	*Loss Tangent*
Alumina	9.4–9.6	0.0001–0.0002
Boron nitride	4.2–4.6	0.0001–0.0003
Beryllia	4.2	0.0005
Borosilicate glass	4.5	0.0008
Pyroceram	5.54–5.65	0.0002
Rayceram	4.70–4.85	0.0002
Slip cast fused silica (SCFS)	3.30–3.42	0.0004
Woven (3D) quartz	3.05–3.1	0.001–0.005
Silicon nitride (HPSN)	7.8–8.0	0.002–0.004
Silicon nitride (RSSN)	5.6	0.0005–0.001
Nitroxyceram	5.2	0.002
Reinforced celsian	6.74	0.0009

Many ceramic materials have a high thermal expansion and are susceptible to thermal shock resulting from rapidly forming temperature gradients during flight. Shock causes the dome to fracture, particularly if the radome is under mechanical loading. For instance, a 400°C differential may result in failure due to thermal shock for an alumina radome [11]. Therefore, the radome designer must take care in material selection.

Slip cast fused silica (SCFS), a form of silicon oxide, is suitable for high-temperature applications. It has good electrical properties, low cost, and a thermal expansion coefficient less than either alumina or cordierite materials, which improves its thermal shock capability. It has been used on reentry vehicles at over Mach 8 velocities.

A high-temperature resin has recently been developed for use in quartz-based 3D composites and in chopped fiber-filled molding compounds [18]. This 3D quartz has good thermal shock resistance and has application in hypersonic missiles or reentry antenna windows. Structures fabricated from this material have been tested at over 1900°C with negligible change in transmission.

A radome designer must be concerned by the variation of a dielectric constant and loss tangent with increasing temperatures because the loss tangent of the material must be kept low to achieve high radome transmissivity. Figure 4.3 shows representative data for SCFS, Pyroceram, and Rayceram. As with dielectric constant, the loss tangent also increases with increasing temperature. Figure 4.4 is a photo of thin wall ceramic radome structures.

Researchers at the Georgia Institute of Technology have evaluated silicon nitride [19] and found it to have good electrical properties, mechanical strength, and high rain erosion and thermal shock resistance. They have investigated it in two forms:

1. Reaction sintered silicon nitride (RSSN) is formed by reaction of silicon with nitrogen in which a flame spray molds silicon powder to shape.
2. Hot pressed silicon nitride (HPSN) is formed from the reaction of silicon with nitrogen to create a powder that can be hot pressed. HPSN is a more dense material than RSSN.

Although silicon nitride generated great interest during the 1980s, to date, only very small radomes have been produced, with some fabrication problems still unresolved.

Several advanced ceramic composites have recently been researched for applications where high-temperature microwave transparency must be combined with good strength and all-weather capability [20]. These composites are as follows:

1. Nitroxyceram, a family of composite materials fabricated from a mixture of silicon nitride, boron nitride, and silica. These materials are fabricated by hot pressing or hot isostatic processing.
2. Celsian, a high-purity barium aluminosilicate. This material exhibits me-

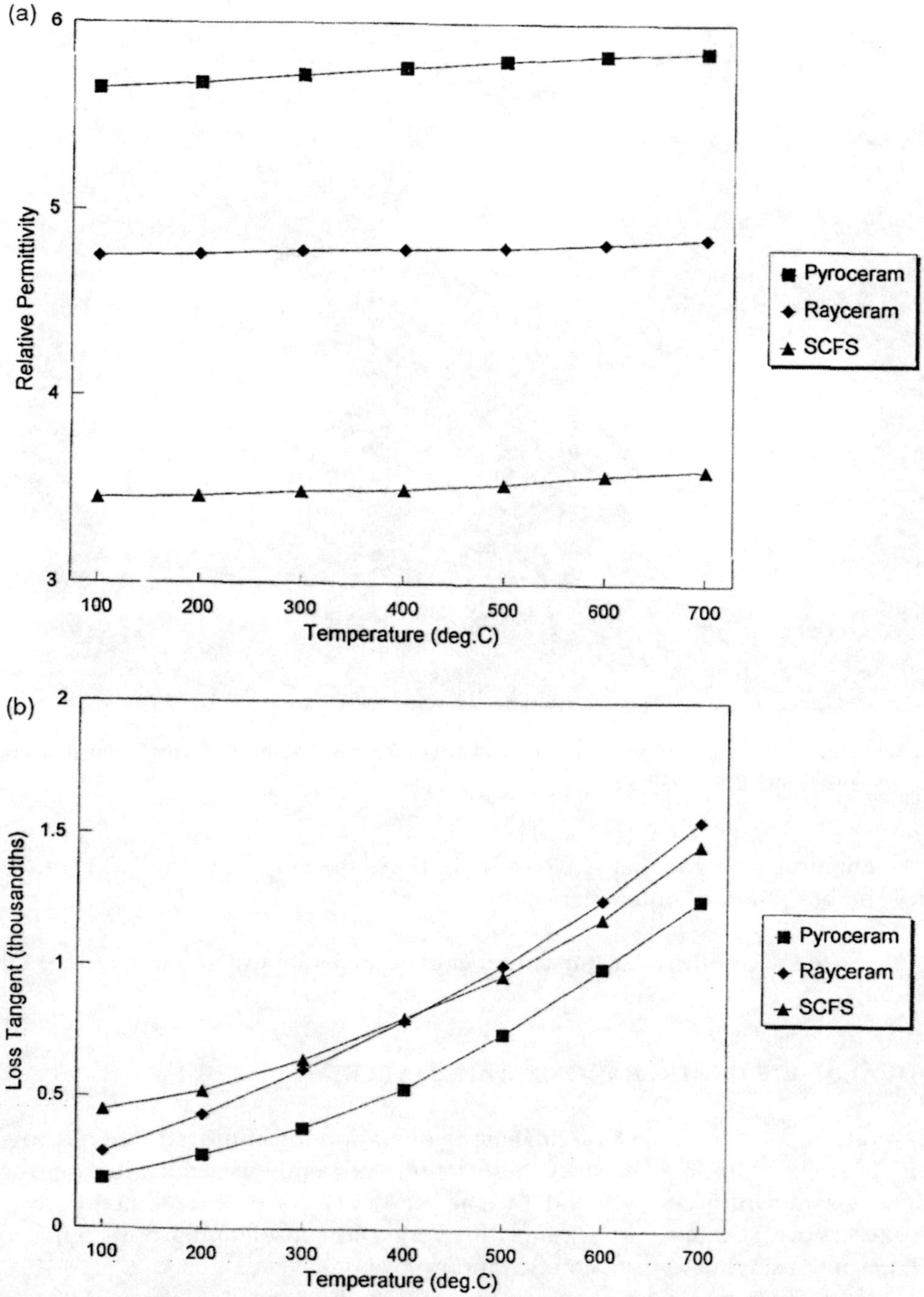

Figure 4.3 Variation with temperature of electrical properties of SCFS, Rayceram, and Pyroceram: (a) relative permittivity and (b) loss tangent.

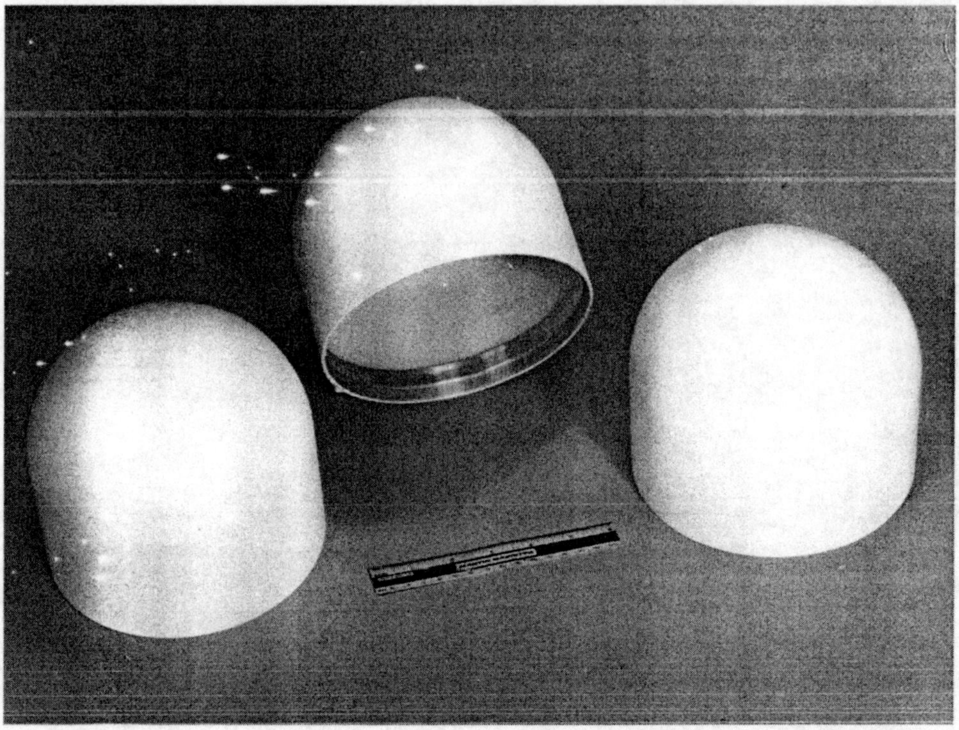

Figure 4.4 Thin wall ceramic radome structures. (Photo courtesy of Lockheed Martin Electronics and Missile Systems Company)

chanical strength that is typically six times greater than silica with much higher rain erosion resistance.

Measured permittivity data for a number of other ceramic materials is published in [21,22].

4.3 DUAL-MODE (RF/IR) RADOME WALL MATERIALS

Many modern military sensors use dual-mode operation for improved target recognition capability. Although most of these sensors currently use individual sensor windows for the infrared (IR) and for the RF (microwave or millimeter-wave) ranges, users are becoming increasingly interested in consolidating both sensors, which permits a single electromagnetic window for both.

Typical IR sensors generally operate in either the near-IR (450- to 1200-μm wavelengths), the mid-IR (3- to 5-μm wavelengths), or the far-IR (8- to 12-μm wavelengths) ranges. These wavelengths generally correspond to atmospheric windows where propagation losses are acceptable.

Nonorganic Dual-Mode Materials

Table 4.5 shows the IR transmittance ranges versus the wavelength for many optical materials. Variations in transmittance of a given material occur for these reasons: (1) manufacturing process differences among suppliers, (2) material impurities, and (3) crystalline structure differences.

The candidates in the list of optical materials become fewer when both of the following requirements for the radome are considered:

1. A small loss tangent, that is, $\tan \delta < 0.01$ in the microwave/millimeter-wave band of interest;

Table 4.5
Transmission Ranges Versus Wavelength of Selected Optical Material

Material	*Wavelengths*
Lithium Floride (LiF)	0.11–6 microns
Magnesim Floride (MgF_2)	0.12–7 microns
Calcium Floride (CaF_2)	0.13–8 microns
Barium Floride (BaF_2)	0.14–12 microns
Quartz (SiO_2)	0.15–3 microns
UV Fused Silica (SiO_2)	0.16–2.5 microns
IR Fused Silica ($SilO_2$)	0.22–3.3 microns
Glass (BK-7)	0.4–1.4 microns
Silicon (Si)	0.45–7 microns
Germanium (Ge)	0.18–12 microns
Zinc Sulfide (ZnS)	0.4–13 microns
Gallium Arsenide(GaAs)	1–14 microns
Sodium Chloride (NaC)	0.18–16 microns
Zinc Selenide (ZnSe)	0.6–28 microns
Potassium Chloride (KCl)	0.19–20 microns
Potassium Bromide (KBr)	0.21–25 microns
Cadmium Telluride (CdTe)	1–25 microns
Silver Chloride (AgCl)	0.4–29 microns
Silver Bromide (AgBr)	0.45–35 microns
Thallium Bromoiodide (KRS-5)	0.6–39 microns
Cesium Bromide (CaBr)	0.22–50 microns
Cesium Iodide (CeI)	0.25–60 microns

2. The thermal and mechanical requirements for flight vehicles, which apply to radomes.

Selection factors are dependent on the application and include strength, hardness, coefficients of expansion, thermal shock resistance, temperature range of operation, and stability of permittivity (or refractive index) with temperature. One of the factors of most concern in terms of currently available optics materials is poor thermal shock resistance.

Researchers at the Georgia Institute of Technology conducted studies of the potential of zinc selenide (ZnSe) and zinc sulfide (ZnS) for optical windows [23]. Texas Instruments (TI) has investigated gallium arsenide (GaAs) as a potential dual-mode material [24]. GaAs is much harder and more resistant to rain erosion than either ZnS or ZnSe. General Electric (GE) has researched chemically deposited silicon nitride as a dual-mode window [25]. GTE Laboratories has also done studies of transparent polycrystalline yttria for IR applications [26]. Domes could be isostatically pressed and sintered, which is an economical fabrication process. Also, GTE Laboratories reported near net shape sintering, which should reduce the cost of grinding and polishing.

The Office of Naval Research (ONR) conducted a study that identified about 30 ceramic materials suitable for dual-mode windows [27]. These include the following:

- Mullite;
- Germanium mullite;
- Boron aluminate;
- Zinc germanate;
- Thorium germanate;
- Pollucite;
- Hafnium titanate;
- Calcium lanthanum sulfide.

In recent years, several new midrange IR materials have become available, including lanthana-strengthened yttria (LSY). The physical and electrical properties of these new materials are reported in [28].

The fabrication process produces LSY in near net shape. The process consists of pressing LSY into the final shape using an isopress mold. The compacted powder is then placed in a high-temperature furnace where it is fired at more than 2100°C. The rest of the process consists of sintering, annealing, and grinding and polishing. At K_a-band, LSY has a relative permittivity of 11.2 and loss tangent of 0.0005. Its high-end infrared cutoff at 7.5 μm is considerably higher than that of other midrange materials. At IR wavelengths, scattering within the crystalline structure is also a major loss factor. Only ceramic windows are possible for the mid- and far-IR bands. However, in the near-IR band, there are many viable organic materials.

Organic Dual-Mode Materials

With organic electromagnetic windows, previous studies have shown that the OH-molecular bond is responsible for severe absorption resonances that fall within the mid- and far-IR bands of interest [29]. These absorption resonances exclude the possibility of using many organic window materials.

However, many plastic materials that provide good radome performance in the microwave and millimeter-wave ranges also offer high transparency in the near-IR range. Nevertheless, compared with ceramics, the lower physical strengths and lower range of potential operating temperatures limit the use of these materials to subsonic aircraft, ground vehicles, or fixed terrestrial applications.

Many organic dual-mode materials exist that give excellent transmission in the near-IR range. These include the following [30]:

- *Polymethyl methacrylate (acrylic)* is the most commonly used plastic and has $\epsilon_r = 2.22$. It is moderately priced, easily molded, and machinable. It has an effective combinatin of transmission, scratch resistance, wavelength stability, and resistance to water absorption. Acrylic materials that transmit in the UV range without noticeable aging are also available.
- *Methylpentene* is optically similar to acrylic, with a relative permittivity of $\epsilon_r = 2.15$. It is tough and has chemical resistance, high temperature resistance, and excellent electrical properties
- *Polystyrene (styrene)* has a relative permittivity of $\epsilon_r = 2.53$. It is a low-cost material with excellent molding properties. Methyl methacrylate styrene (MAS) is a copolymer material consisting of 70% acrylic and 30% styrene. The relative permittivity can vary from $\epsilon_r = 2.35$ to 2.45, depending on the blend. The material can be machined and polished, molds easily, and is considered stable. Styrene acrylonitrile (SAN) is also an acrylic styrene copolymer with $\epsilon_r = 2.5$. The material is thermally stable and molds well, but has a tendency to yellow.
- *Polycarbonate* maintains its physical strength over a temperature range of $-135°$ to $+120°$C, making it suitable for airborne applications. Its relative permittivity is approximately $\epsilon_r = 2.52$. The material is more difficult to mold than either acrylic or styrene, and is not easily machined or polished due to its high ductility. The major advantage is its high impact resistance, which makes it the preferred material where durability is important. Its cost is significantly high among optical plastics.
- *Nylon* has a relative permittivity of $\epsilon_r = 2.4$ to 2.9. It is used in applications requiring a combination of transparency and resistance to a wide range of solvents and chemicals. It displays excellent electrical properties with strength and stiffness. Moisture absorption is high; therefore, dimensional stability is low. Nylon's high-flow characteristics and melt stability allow easy processing and the ability to fill thin-wall sections. It is one of the more expensive optical plastics.

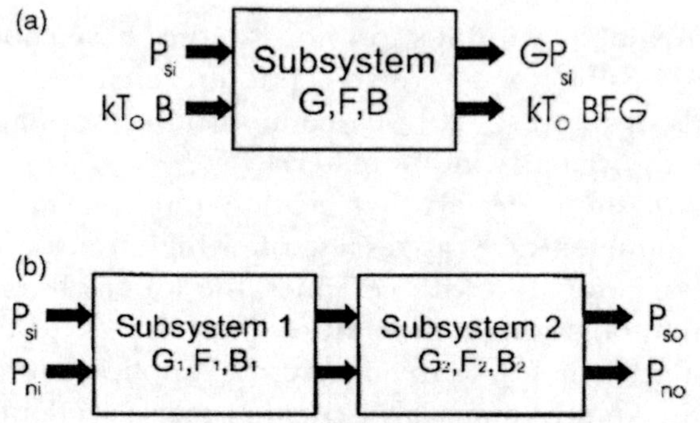

Figure 4.5 Representation of noisy subsystems: (a) subsystem model and (b) two subsystems in cascade.

4.4 EFFECT OF RADOME MATERIAL ON ANTENNA NOISE TEMPERATURE

Radome losses on the antenna noise temperature decrease receiver sensitivity, particularly in a radome-enclosed satellite antenna receiving system. For purposes of subsystems signal-to- noise calculations, it is customary to model the system as noiseless, but account for the noise at the output by a noise factor NF, as illustrated in Figure 4.5(a). The quantity $k_b T_0 B_s$ is the actual thermally generated noise power at the input port to a subsystem. The noise factor is a dimensionless quantity. Using the well-known laws of blackbody radiation [31], the additional noise introduced by a network and referenced to its output port is

$$P_{no} = k_b T_0 B_s G_s (NF - 1) \tag{4.2}$$

where

G_s = Subsystem power gain (dimensionless);
B_s = Noise bandwidth of the system (Hz);
k_b = Boltzmann's constant = 1.38×10^{-23} W-s/K;
T_0 = Ambient temperature = 290K.

The noise figure is the noise factor expressed in decibels:

$$F = 10 \log (NF) \tag{4.3}$$

Alternatively, you can account for subsystem-generated noise via a noise-free termination having a physical temperature T_e. This equivalent noise temperature would produce the same noise power at the output. The effective noise temperature may be expressed in terms of the system noise factor as in [32]:

$$T_e = (NF - 1)T_0 \tag{4.4}$$

Note that this expression only applies for a particular terminating impedance at the receiver input. If another value is used, the noise temperature changes. If you specify a receiver noise temperature without a quoted impedance specification, presume that the optimum source impedance is used. For instance, many receiver systems use 50-Ω coaxial transmission lines, and they are matched into a 50-Ω receiver load impedance.

Noise factor and equivalent noise temperature describe the same property of the subsystem. However, noise temperature is the preferred method to describe low-noise receiving systems.

Figure 4.5(b) shows the ideas describing noise of individual subsystems applied to cascaded subsystems. The noise factor and the noise temperature for the cascade systems are given by the parameters for the individual subsystem via

$$F = NF_1 + \frac{NF_2 - 1}{G_1} \tag{4.5}$$

$$T = T_1 + \frac{T_2}{G_1} \tag{4.6}$$

From these relationships, note that if the gain of the first stage is sufficiently high, the first stage essentially sets the overall system noise performance. This is the motivation for cascaded subsystems, and is particularly relevant in low-noise receiving systems where the first stage is a low-noise amplifier (LNA) and the second stage the microwave receiver. The state of the art in C-band satellite television receive-only (TVRO) LNAs is of the order of 25K.

Placed over an antenna, a radome having a power loss L can seriously degrade system noise temperature. Consider the receiver front-end illustrated in Figure 4.6. In this model, derive the antenna noise temperature T_a from the antenna's radiation pattern and its surrounding environment. A zenith-pointing satellite antenna may have antenna temperatures of the order of 20K due to the low sky temperatures. The system noise temperature, referenced to the receiver input port, is given by

$$T = T_{\text{LNA}} + (1 - \eta_a)T_0 + \eta_a[T_a(1 - L) + T_r L] \tag{4.7}$$

where

T_{LNA} = Noise temperature of LNA;
η_a = Antenna radiation efficiency ($0 \leq \eta_a \leq 1$);

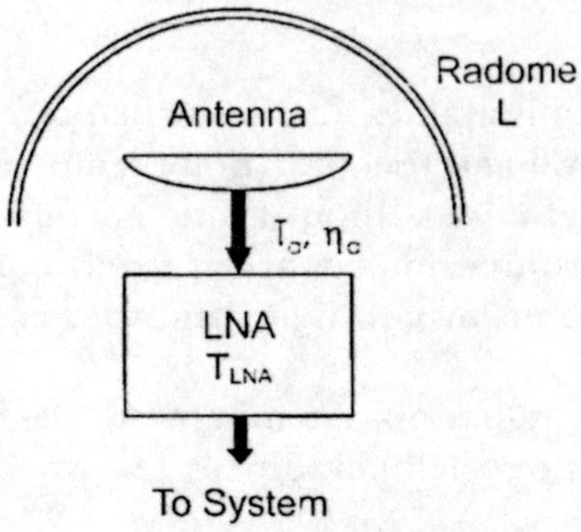

Figure 4.6 Satellite receiving antenna enclosed in a radome of loss L.

T_r = Physical temperature of radome (K);
L = Radome power transmission loss factor ($0 \leqq L \leqq 1$).

Note that the radome power loss factor is related to the radome loss in decibels via the formula

$$L_{\text{dB}} = 10 \log (L) \tag{4.8}$$

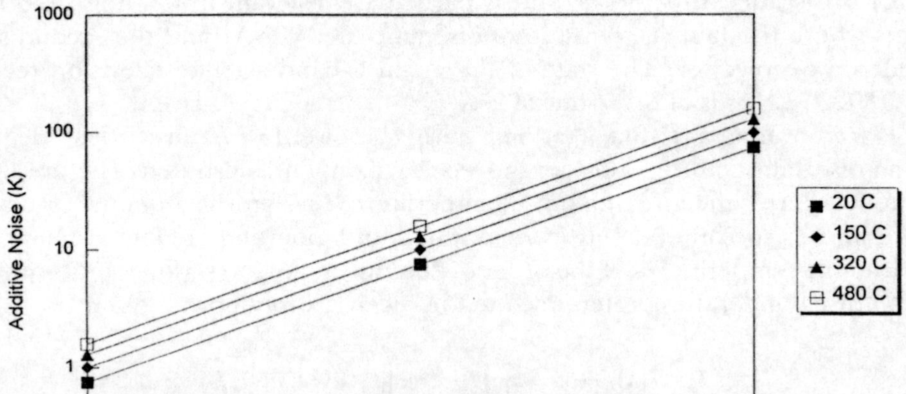

Figure 4.7 Additive system noise due to a heated lossy radome.

Figure 4.8 Bulge on upper portion of UAV fuselage is a radome for a SATCOM antenna. (Photo courtesy of General Atomics Aeronautical Systems, Inc.)

For a low-noise system, Figure 4.7 shows the additive system noise temperature due to radome loss and physical temperature. These data show that at ambient temperatures, the radome loss can be a large contributor to system noise temperature. For instance, every −0.1 dB of radome loss can degrade system noise temperature of the order of 10 K. Also, note that missile radomes operating at very high physical temperatures may further degrade the sensitivity of radome-enclosed receiving systems. Radomes wetted by rain have also been found to exhibit increased system noise temperature degradation [33]. To minimize this effect, Teflon was investigated as a hydrophobic coating [34]. The effectiveness of various commercially available hydrophobic coatings is listed in [28]. Figure 4.8 shows a SATCOM radome installed on an unmanned aerial vehicle (UAV).

References

[1] Bassett, H. L., and J. M. Newton, Eds., *Proceedings of the 13th Symposium on Electromagnetic Windows*, Georgia Institute of Technology, Atlanta, GA, September 1976.

[2] Harris, J. N., Ed., *Proceedings of the 14th Symposium on Electromagnetic Windows*, Georgia Institute of Technology, Atlanta, GA, June 1978.

[3] Bassett, H. L., and G. K. Huddleston, Eds., *Proceedings of the 15th Symposium on Electromagnetic Windows*, Georgia Institute of Technology, Atlanta, GA, June 1980.

[4] Huddleston, G. K., Ed., *Proceedings of the 16th Symposium on Electromagnetic Windows*, Georgia Institute of Technology, Atlanta, GA, June 1982.

[5] Bassett, H. L., Ed., *Proceedings of the 17th Symposium on Electromagnetic Windows*, Georgia Institute of Technology, Atlanta, GA, July 1984.

[6] Harris, J. N., Ed., *Proceedings of the 18th Symposium on Electromagnetic Windows*, Georgia Institute of Technology, Atlanta, GA, September 1986.

[7] Handley, J. C., Ed., *Proceedings of the 19th Symposium on Electromagnetic Windows,* Georgia Institute of Technology, Atlanta, GA, September 1988.

[8] Ohlinger, W. L., Ed., *Proceedings of the 20th Symposium on Electromagnetic Windows,* Georgia Institute of Technology, Atlanta, GA, September 1992.

[9] Handley, J. C., Ed., *Proceedings of the 21st Symposium on Electromagnetic Windows,* Georgia Institute of Technology, Atlanta, GA, January 1995.

[10] Cary, R. H., "Avionics Radome Materials," North Atlantic Treaty Organization Report AGARD-AR-75, London, Technical Editing and Reproduction Ltd., 1974.

[11] Walton, J. D., *Radome Engineering Handbook: Design and Principles, Ceramics and Glass,* Science and Technology Series, New York: Marcel Dekker, 1970.

[12] Ossin, A., et al., "Millimeter Wavelength Radomes," AFML-TR-79-4076, Air Force Materials Laboratory, Wright Patterson Air Force Base, OH, 1979.

[13] Gibson, C. C., R. H. Bogaard, and D. L. Taylor, "The EM Window Material Database: A Progress Report," *Proc. 21st Symp. on Electromagnetic Windows,* Georgia Institute of Technology, Atlanta, GA, January 1995.

[14] Skolnik, M. I., *Radar Handbook,* New York: McGraw Hill Book Company, 1970.

[15] Newton, J. M., D. J. Kozakoff, and J. M. Schuchardt, "Methods of Dielectric Material Characterization at Millimeter Wavelengths," *Proc. 15th Symp. on Electromagnetic Windows,* Georgia Institute of Technology, Atlanta, GA, June 1980.

[16] Rudge, A. W., K. Milne, A. D. Olver, and P. Knight, *The Handbook of Antenna Design,* IEEE Electromagnetic Waves Series, London, Peter Peregrinus Ltd., 1986.

[17] Coy, T. N., "Hot Pressed Silicon Nitride," *Proc. 11th Symp. on Electromagnetics Windows,* Georgia Institute of Technology, Atlanta, GA, 1972.

[18] Favaloro, M. R., "A Dielectric Composite Material for Advanced Radome and Antenna Window Applications," *Proc. 20th Symp. on Electromagnetic Windows,* Georgia Institute of Technology, Atlanta, GA, September 1992.

[19] Walton, J.D., "Reaction Sintered Silicon Nitride," *Proc. 11th Symp. on Electromagnetics Windows,* Georgia Institute of Technology, Atlanta, GA, 1972.

[20] Wright, J. M., J. F. Meyers, and E. E. Ritchie, "Advanced Ceramic Composites for Hypersonic Radome Applications," *Proc. 20th Symp. on Electromagnetic Windows,* Georgia Institute of Technology, Atlanta, GA, September 1992.

[21] Sheppard, A. P., A. McSweeney, and K. H. Breeden, "Submillimeter Wave Material Properties and Techniques," *Proc. Symp. on Submillimeter Waves,* New York, Polytechnic Institute of Brooklyn, 1970.

[22] Nickel, H. U., and R. Heidinger, "A Survey of Vacuum Windows for High Energy Millimeter Wave Systems in Fusion Experiments," *Proc. 20th Symp. on Electromagnetic Windows,* Georgia Institute of Technology, Atlanta, GA, September 1992.

[23] Papis, J., B. DiBeneditto, and A. Swanson, "Low Cost Zinc Selenide and Zinc Sulfide for FLIR Optics," *Proc. 13th Symp. on Electromagnetic Windows,* Georgia Institute of Technology, Atlanta, GA, 1976.

[24] D. Purinton, "GaAs IR-RF Window Material," *Proc. 13th Symp. on Electromagnetic Windows,* Georgia Institute of Technology, Atlanta, GA, 1976.

[25] Tanzilli, R. A., J. J. Gebhardt, and J. O. Hanson, "Potential of Chemically Vapor Deposited Silicon Nitride as a Multimode EM Window," *Proc. 14th Symp. on Electromagnetic Windows,* Georgia Institute of Technology, Atlanta, GA, 1978.

[26] Rhodes, W. H., "Transparent Polycrystalline Yttria for IR Applications," *Proc. 16th Symp. on Electromagnetic Windows,* Georgia Institute of Technology, Atlanta, GA, 1982.

[27] Musikant, S., et al., "Advanced Optical Ceramics," *Proc. 15th Symp. on Electromagnetic Windows,* Georgia Institute of Technology, Atlanta, GA, 1976.

[28] Rhodes, W. H., M. R. Pascucci, G. C. Wei, E. A. Trickett, and R. A. Long, "Lanthana Strengthened Yttria as a Mid-range Infrared Window Material," *Proc. 20th Symp. on Electromagnetic Windows,* Georgia Institute of Technology, Atlanta, GA, September 1992.

[29] Townes, C. H., and A. L. Shawlow, *Microwave Spectroscopy,* New York: McGraw Hill Book Company, 1965.

[30] Tribastone, C., and C. Teyssier, "Designing Plastic Optics for Manufacturing," *Photonics Spectra Magazine,* Laurin Publishing Company, January 1991.

[31] Skolnik, M.I., *Introduction to Radar Systems,* 2nd ed., New York: McGraw Hill Book Company, 1980.

[32] Ulaby, F. T., R. K. Moore, and A. K. Fung, *Microwave Remote Sensing: Vol. 1,* Norwood, MA: Artech House, 1981.

[33] Dijk, J., and A. C. A. Van der Vorst, "Depolarization and Noise Properties of Wet Antenna Radomes," *AGARD Proc.,* No.159, Paris, France, October 1974.

[34] Siller, C. A., "Preliminary Testing of Teflon as a Hydrophobic Coating for Microwave Radomes," *IEEE Trans. Antennas Propagation,* Vol. AP-27, No. 4, July 1979.

PART II

Radome Analytical Techniques

Chapter 5

Multilayer Dielectric Walls

The aim of this chapter is to introduce the reader to the most common radome wall constructions. The term *sandwich* when applied to radomes covers any multilayer construction in which the layers differ in dielectric constant. All of these can be mathematically modeled by the boundary value problem solution presented in Section 5.1. Other less common radome walls such as serrated or grooved walls [1,2] are not treated here.

5.1 MATHEMATICAL FORMULATION FOR RADOME WALL TRANSMISSION

The analysis of microwave transmission and reflection for a flat multilayer lamination can be achieved via a boundary value solution of the *N*-layer dielectric wall depicted in Figure 5.1 for forward and reverse propagating waves (C_i and B_i, respectively), where the solution takes the form [3–5]:

$$\begin{bmatrix} C_1 \\ B_1 \end{bmatrix} = \prod_{i=1}^{N} \frac{1}{T_i T_{N-1}} \begin{bmatrix} e^{j\gamma_i t_i} & R_i e^{-j\gamma_i t_i} \\ R_i e^{j\gamma_i t_i} & e^{-j\gamma_i t_i} \end{bmatrix} \begin{bmatrix} 1 & R_{N+1} \\ R_{N+1} & 1 \end{bmatrix} \begin{bmatrix} C_{N+1} \\ B_{N+1} \end{bmatrix} \quad (5.1)$$

where

t_i = Layer thickness of i'th layer;
R_i, T_i = Fresnel reflection and transmission coefficients, respectively, of the i'th layer free space;
γ_i = Propagation constant in the i'th layer.

In terms of the angle-of-incidence relative to the radome normal, θ,

$$\gamma_i = k_0 \sqrt{\epsilon_{ri} - \sin^2 \theta} \quad (5.2)$$

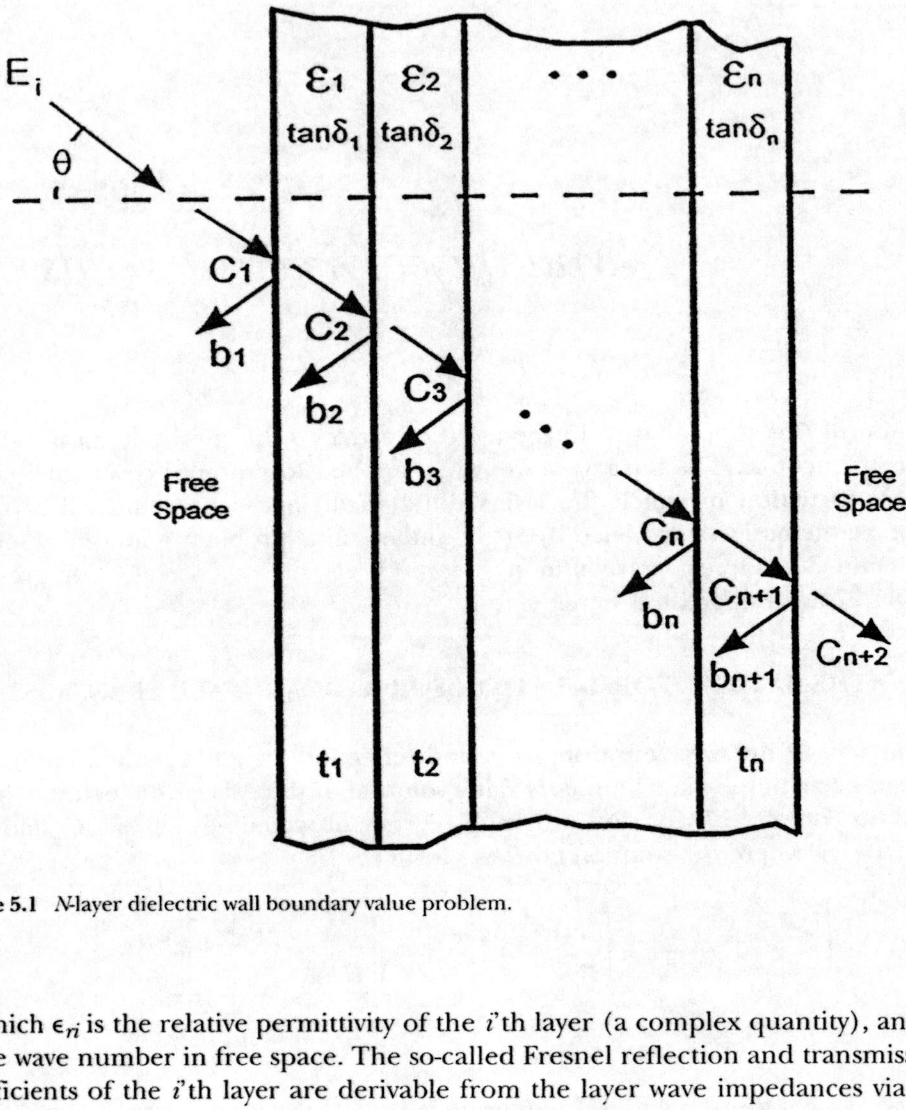

Figure 5.1 *N*-layer dielectric wall boundary value problem.

in which ϵ_{ri} is the relative permittivity of the i'th layer (a complex quantity), and k_0 is the wave number in free space. The so-called Fresnel reflection and transmission coefficients of the i'th layer are derivable from the layer wave impedances via the following equations:

$$R_i = \frac{Z_i - Z_{i-1}}{Z_i + Z_{i-1}}; \; T_i = 1 - R_i \tag{5.3}$$

The result of the matrix computation depends on the polarization of the incident electromagnetic wavefront. In these expressions, the user must evaluate the layer impedances for perpendicular or parallel polarization. For perpendicular polarization,

$$Z_i = \frac{\cos\theta}{\sqrt{\epsilon_{ri} - \sin^2\theta}} \tag{5.4}$$

and for parallel polarization

$$Z_i = \frac{\sqrt{\epsilon_{ri} - \sin^2\theta}}{\epsilon_{ri}\cos\theta} \tag{5.5}$$

Note that the dielectric loss is accounted for by a loss tangent term for the dielectric material:

$$\epsilon_{ri} = \epsilon'_{ri}(1 - j\tan\delta_i) \tag{5.6}$$

where ϵ'_{ri} is the real part of the relative permittivity for the i'th layer. Alternatively, one could modify the preceding equations to include a conductivity term.

After multiplying out the matrix product, the result is obtained:

$$\begin{bmatrix} C_1 \\ B_1 \end{bmatrix} = \begin{bmatrix} A_{11} & A_{12} \\ A_{21} & A_{22} \end{bmatrix} \begin{bmatrix} C_{N+2} \\ B_{N+2} \end{bmatrix} \tag{5.7}$$

The front face voltage reflection coefficient from the sandwich wall is a complex number:

$$R_w = \frac{B_1}{C_1}\bigg|_{(B_{N+2}=0)} = \frac{A_{21}}{A_{11}} \tag{5.8}$$

Likewise, the transmission voltage coefficient is

$$T_w = \frac{C_{N+2}}{C_1}\bigg|_{(B_{N+2}=0)} = \frac{1}{A_{11}} \tag{5.9}$$

The transmission coefficient is also complex and can be expressed in terms of its magnitude and the insertion phase delay (IPD) angle:

$$T_W = |T_W| \angle \text{IPD} \tag{5.10}$$

Engineers working in radome design often use the term *radome transmission efficiency* for

$$|T_w|^2 \tag{5.11}$$

In Appendix 5A at the end of this chapter, a computer software listing is provided for the preceding matrix solution for multilayer wall transmission coefficient. The software coding was accomplished in Microsoft BASIC.

Transmission Coefficients for Circular Polarized Waves

Transmission coefficients for circular polarized waves are computed from the following formulas:

$$T_{W\text{copol}} = \frac{T_{W\perp} - T_{W\|} \sin \Delta_{\text{IPD}} + T_{W\perp} \cos \Delta_{\text{IPD}}}{2} \tag{5.12}$$

$$T_{W\text{xpol}} = \frac{T_{W\perp} - T_{W\|} \sin \Delta_{\text{IPD}} - T_{W\perp} \cos \Delta_{\text{IPD}}}{2} \tag{5.13}$$

in which

$T_{W\|}, T_{W\perp}$ = Parallel and perpendicular polarization transmission coefficients, respectively;

Δ_{IPD} = $(\text{IPD}_{\perp} - \text{IPD}_{\|})$;

$\text{IPD}_{\perp}, \text{IPD}_{\|}$ = Insertion phase delays associated with the perpendicular and parallel polarization transmission coefficients, respectively.

These copolarized and cross-polarized transmission coefficients are useful to calculate the effects of radome depolarization on antenna performance.

5.2 RADOME TYPES, CLASSES, AND STYLE DEFINITIONS

Radomes for use on flight vehicles (aircraft or missiles), surface vehicles, and fixed ground installations are classified into various categories according to MIL-R-7705B [6]. These categories are determined by the specific radome use and wall construction.

There are six types of radomes as identified in the following list.

- *Type I:* Low-frequency radomes used at frequencies at or below 2.0 GHz.
- *Type II:* Directional guidance radomes having specified directional accuracy requirements. These include boresight error (BSE), boresight error slope (BSES), antenna pattern distortion, and antenna sidelobe degradation.
- *Type III:* Narrowband radomes with an operational bandwidth of less than 10%.
- *Type IV:* Multiple-frequency band radomes used at two or more narrow–frequency bands.
- *Type V:* Broadband radomes generally providing an operational bandwidth between 0.100 and 0.667 GHz.
- *Type VI:* Very broadband radomes that provide an operational bandwidth greater than 0.667 GHz.

In addition to the type, the referenced MIL specification defines classes of radomes according to the general application: Class I (flight vehicles), Class II (surface vehicles), and Class III (fixed ground installations).

Radome styles are defined according to their dielectric wall construction. The five basic styles are described in the following list:

- *Style a:* Radomes are half-wave wall solid (monolithic) structures.
- *Style b:* Radomes are thin-wall monolithic structures with a wall thickness equal to or less than 0.1λ wavelengths at the highest operating frequency.
- *Style c:* Radomes are so-called A-sandwich multilayer walls. These consist of three layers: two high-density skins and a low-density core material. The dielectric constant of the skins is greater than the dielectric constant of the core material.
- *Style d:* Radomes are multilayer walled structures having five or more dielectric layers. These have an odd number of high-density skin layers and an even number of low-density core layers. The dielectric constants of the skins are greater than the dielectric constants of the core materials. As the number of layers increases, the broadband frequency performance improves.
- *Style e:* Radomes of this style are defined as all possible radome wall constructions not fitting into Styles a through d. These include the B-sandwich type and others. The B-sandwich is similar to the A-sandwich except it consists of two low-density skins and a high-density core material. The dielectric constant of the skins is less than the dielectric constant of the core material.

5.3 WALL STYLE ELECTRICAL PERFORMANCE

This section examines design criteria and typical transmission performance of the various radome wall styles previously defined. Some of the most commonly used radome walls are illustrated in Figure 5.2 and a high-performance monolithic radome is shown in Figure 5.3.

Figure 5.2 Illustration of the most common sandwich radome walls.

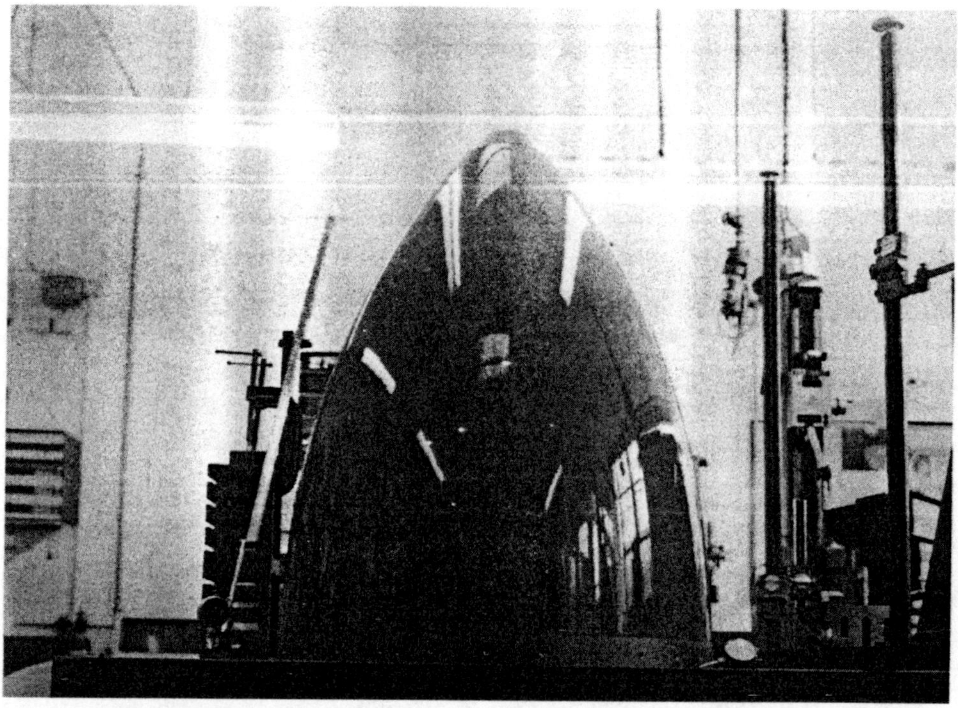

Figure 5.3 High-performance monolithic radome for Hawk 200 fighter. (Photograph courtesy of British Aerospace.)

Half-Wave Wall Radomes (Style a)

The characteristics of a half-wave wall are dependent on the interactions at the two air/dielectric interfaces. They are inherently narrowbanded with typically 5% bandwidth achievable and having performance specifications consistent with normal radome applications. Wall thicknesses that are multiples of a half-wavelength have even narrower operational bandwidth. Another disadvantage of monolithic half-wave walled radomes is weight.

For the half-wave wall radome, the wall thickness for lowest transmission loss is a function of incidence angle and wavelength of the electromagnetic radiation, and the wall material dielectric constant. This relationship can be expressed as follows:

$$t = \frac{m\lambda}{2\sqrt{\epsilon_r - \sin^2\theta}} \tag{5.14}$$

Here, λ is the wavelength of radiation and m the wall order (an integer $\geqq 1$).

The voltage transmission coefficient for a monolithic half-wave wall with a thickness of 0.8255 cm is shown in Figures 5.4 and 5.5 for perpendicular and parallel

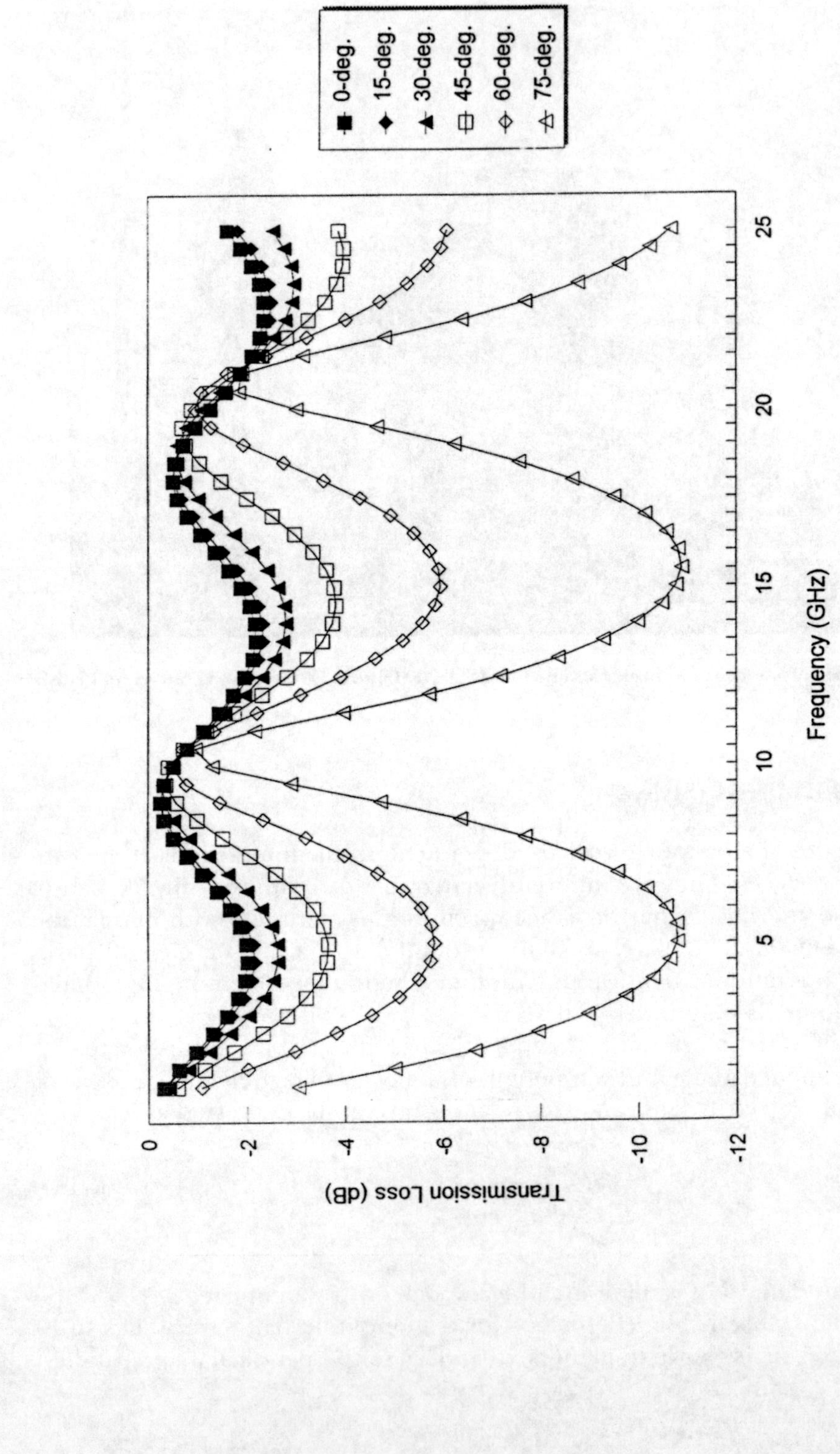

Figure 5.4 Half-wave wall transmission with perpendicular polarization.

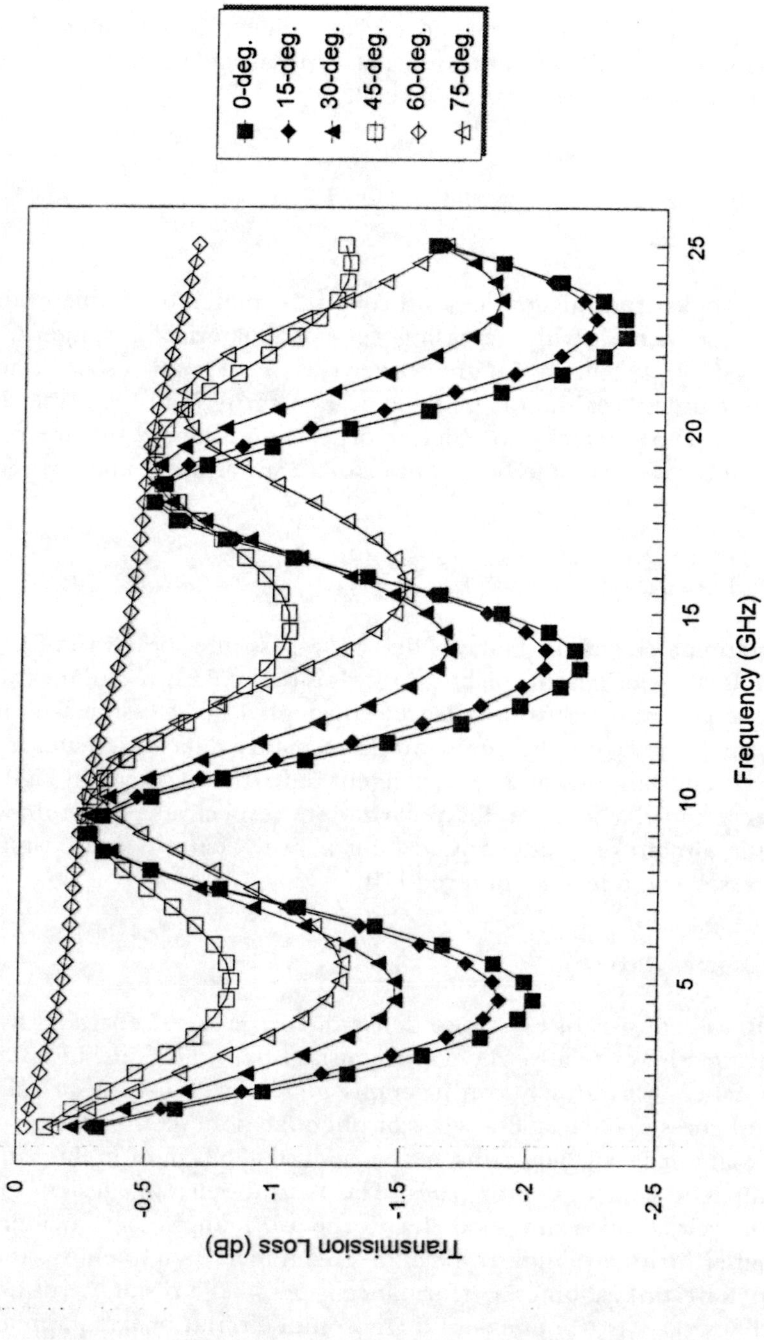

Figure 5.5 Half-wave wall transmission with parallel polarization.

polarization, respectively. The material had a relative dielectric constant of 4 and loss tangent of 0.015. The parallel polarization data, in particular, demonstrates zero front face reflection at the Brewster angle given by

$$\theta_B = \tan^{-1}\left(\sqrt{\epsilon_r}\right) \tag{5.15}$$

At the Brewster angle, transmission losses through the material are due entirely to ohmic losses. These increase with increasing value of a material loss tangent.

The use of multiple half-wave higher order (i.e., $m > 1$) wall designs is not normally considered unless realization of structural integrity is not possible by any other means [7]. The penalties for higher order monolithic walls are reduced bandwidth and increased radome boresight error (BSE) and radome boresight error slope (BSES).

Thin-Walled Radomes (Style b)

The thin-wall radome essentially behaves like a low-pass filter below the frequency for which it is 0.05 wavelengths in thickness [8]. It is particularly useful for transmission incidence angles ranging from 0 degrees through 70 degrees measured from the surface normal [9]. Typical transmission performance data for a material with a relative dielectric constant of 4 and a loss tangent of 0.015 are shown in Figures 5.6 and 5.7 for perpendicular and parallel polarization, respectively. The thin-wall radome finds little aircraft or missile applications above X-band because significant mechanical stresses would be encountered [10].

A-Sandwich Radomes (Style c)

An A-sandwich wall consists of two more dense dielectric skins separated by a less dense core having a lower relative dielectric constant than the skins. In aircraft radomes, for instance, A-sandwich constructions generally utilize electrically thin plastic skins and cores made of a fiberglass or phenolic honeycomb, or else a lightweight plastic foam material. The skins are normally 0.0762 cm or greater in order to provide sufficient structural properties. The A-sandwich wall designs provide high strength-to-weight ratios and good electrical performance at low incidence angles. At the higher incidence angles typical of streamlined high-fineness-ratio (i.e., length-to-diameter ratio) radomes, performance is poor. As a result, A-sandwich radomes are only used on low-fineness-ratio (subsonic) aircraft or fixed applications where near hemispherical shapes may be utilized. These radomes present low incidence angles.

To demonstrate typical A-sandwich performance, consider a sandwich with 0.0762-cm quartz polycyanate skins ($\epsilon_r = 3.23$, $\tan\delta = 0.016$) and a 1.016-cm phenolic

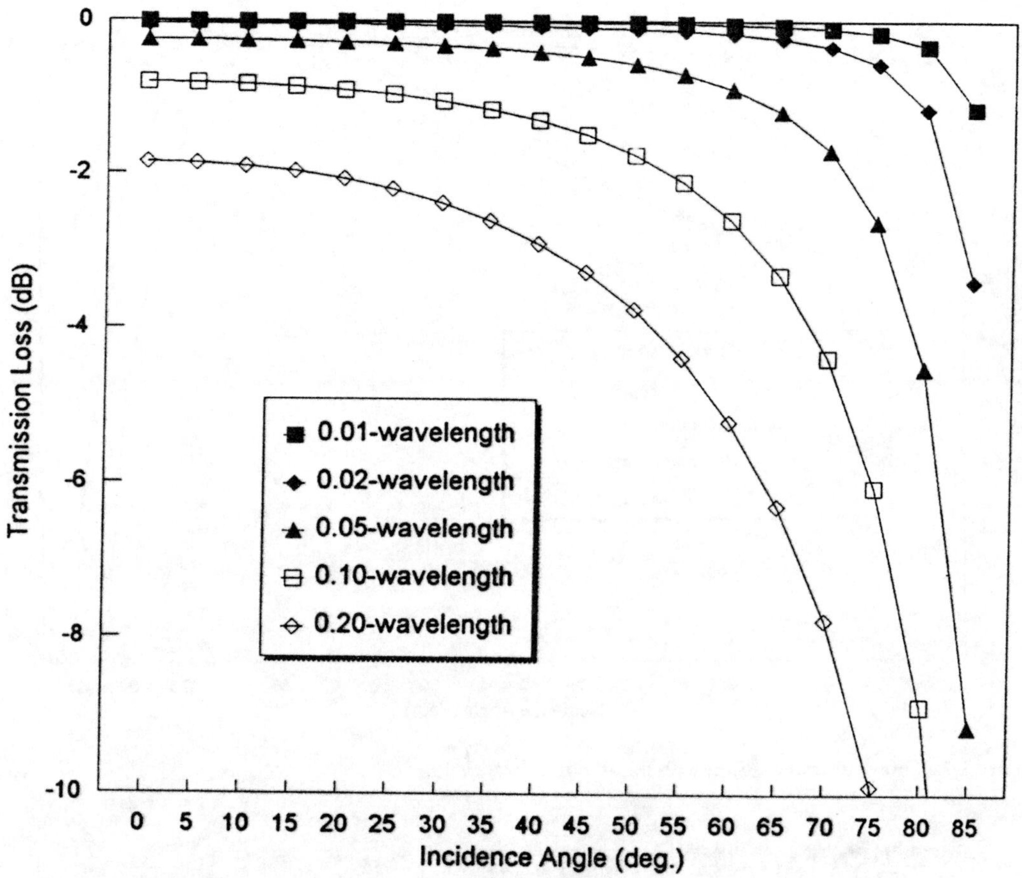

Figure 5.6 Thin-wall transmission with perpendicular polarization.

honeycomb core ($\epsilon_r = 1.10$, $\tan \delta = 0.001$). The wall transmission versus frequency data are shown for perpendicular and parallel polarization in Figures 5.8 and 5.9, respectively.

Multilayered Wall Radomes (Style d)

Multilayer or sandwich radome wall designs having five or more layers were developed in answer to the need for increased operating bandwidth and reduced weight. Sandwich walls are more flexible in design than monolithic walls because of their greater degrees of freedom. For instance, multiple frequency or broadband operation is more easily accommodated. The primary disadvantage is that at higher angles-of-incidence, sandwich wall radomes have greater insertion phase delays (IPDs) than monolithic walled radomes. This results in higher BSE and BSES values.

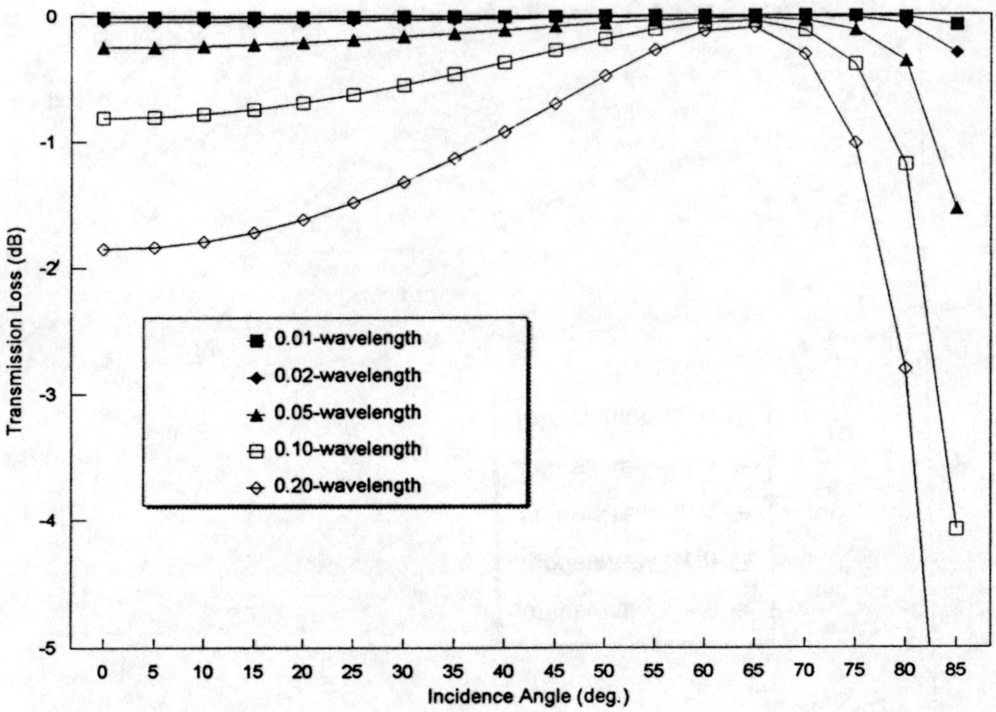

Figure 5.7 Thin-wall transmission with parallel polarization.

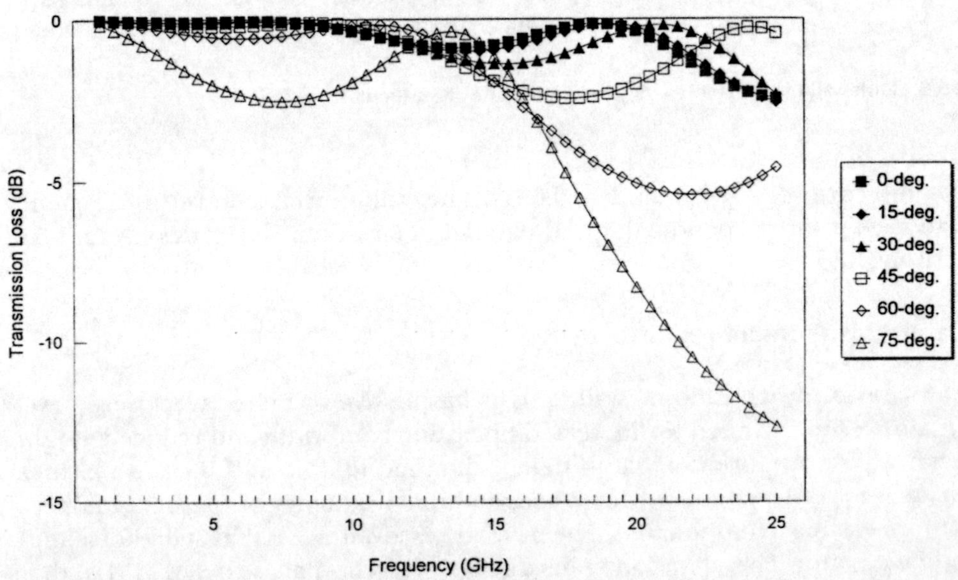

Figure 5.8 A-sandwich transmission with perpendicular polarization.

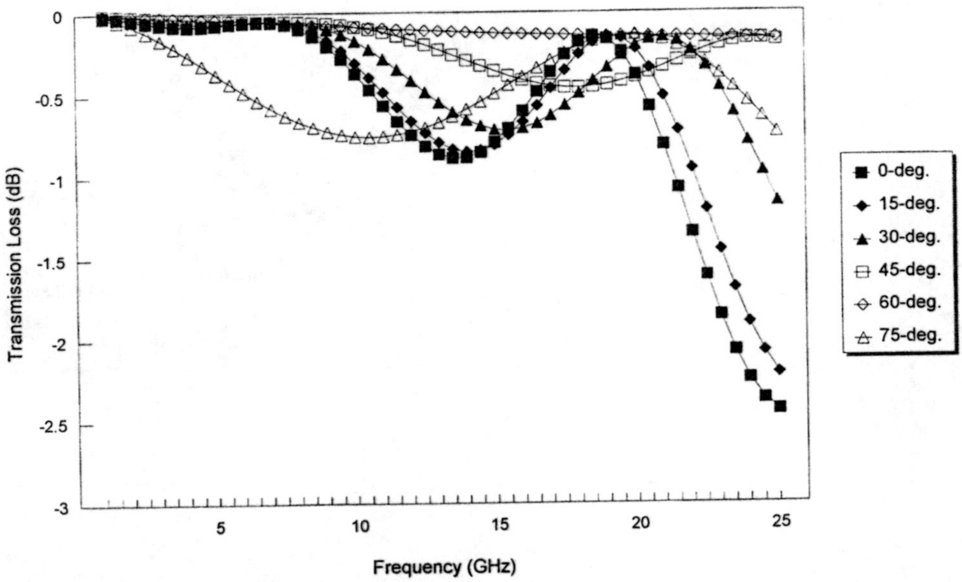

Figure 5.9 A-sandwich transmission with parallel polarization.

The so-called C-sandwich wall consists of a double A-sandwich design. It offers greater operating bandwidth than the A-sandwich. Also, when properly designed, the C-sandwich provides reasonably good performance even at elevated incidence angles. A representative C-sandwich with 0.0762-cm quartz polycyanate skins and 0.5842-cm phenolic honeycomb cores was considered. Computed transmission data versus frequency are shown in Figures 5.10 and 5.11 for perpendicular and parallel polarization, respectively.

Increasing the number of layers most often increases the transmission bandwidth. For instance, the five-layer (C-sandwich) wall design was modified to a seven-layer system. Here, the quartz polycyanate skins are 0.0762 cm but the phenolic honeycomb cores were reduced to 0.3302 cm. The computed transmission versus frequency data are shown in Figures 5.12 and 5.13 for perpendicular and parallel polarization, respectively. Comparing these data to the previous C-sandwich data, considerable improvement in transmission bandwidth is observed.

B-Sandwich Walled Radomes (Style e)

A B-sandwich wall is similar to the A-sandwich wall, but it has less dense dielectric skins separated by a more dense (higher dielectric constant) core. The use of B-sandwich wall radomes has been primarily constrained to terrestrial applications as opposed to flight vehicles because of structural problems associated with low density dielectric skins such as rain erosion problems.

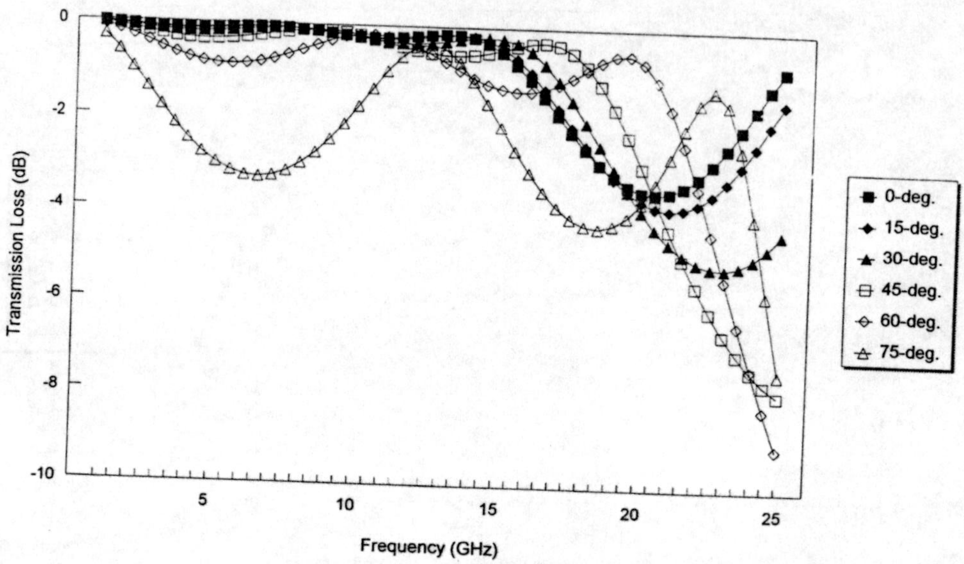

Figure 5.10 C-sandwich wall transmission with perpendicular polarization.

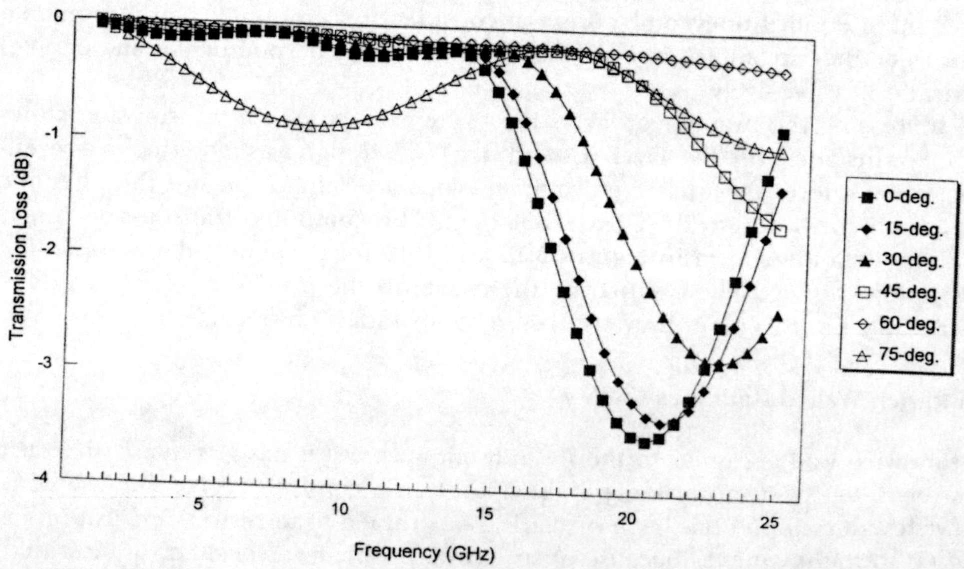

Figure 5.11 C-sandwich wall transmission with parallel polarization.

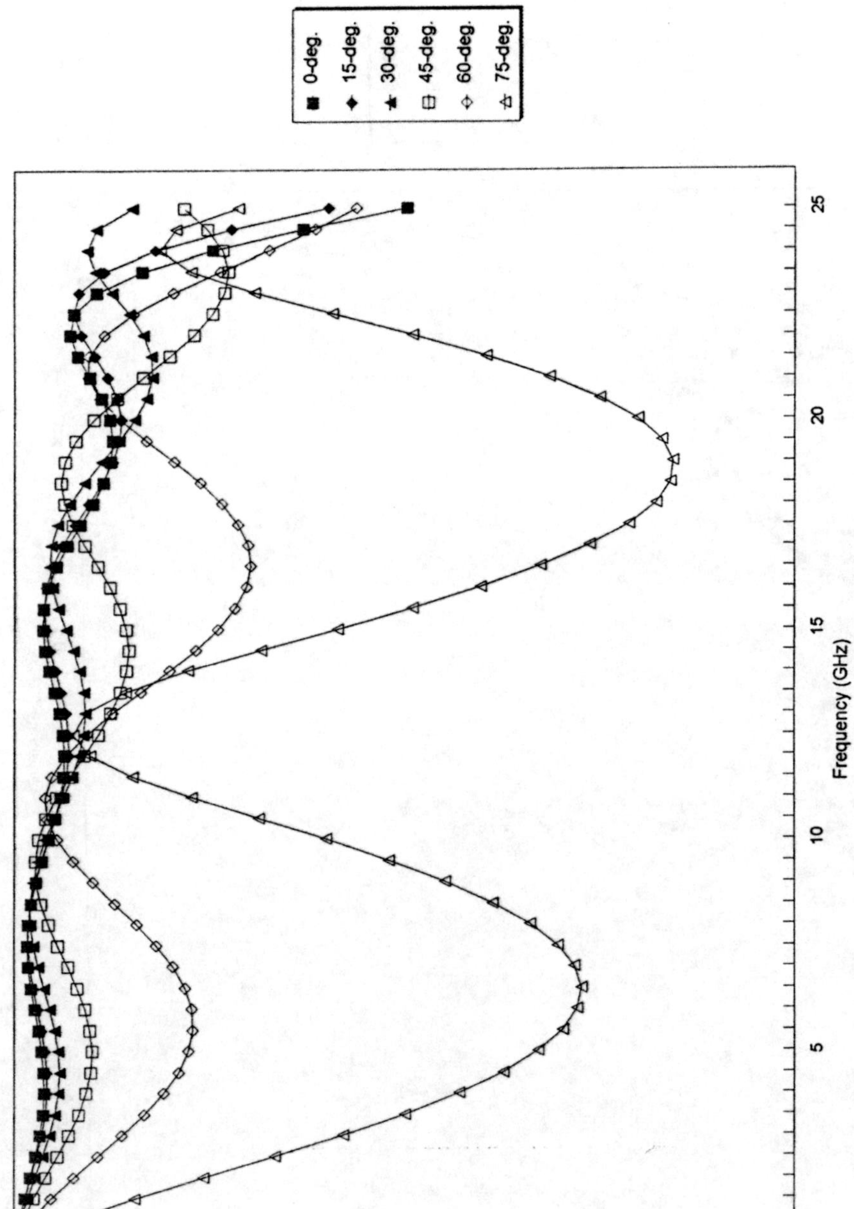

Figure 5.12 Seven-layer wall transmission with perpendicular polarization.

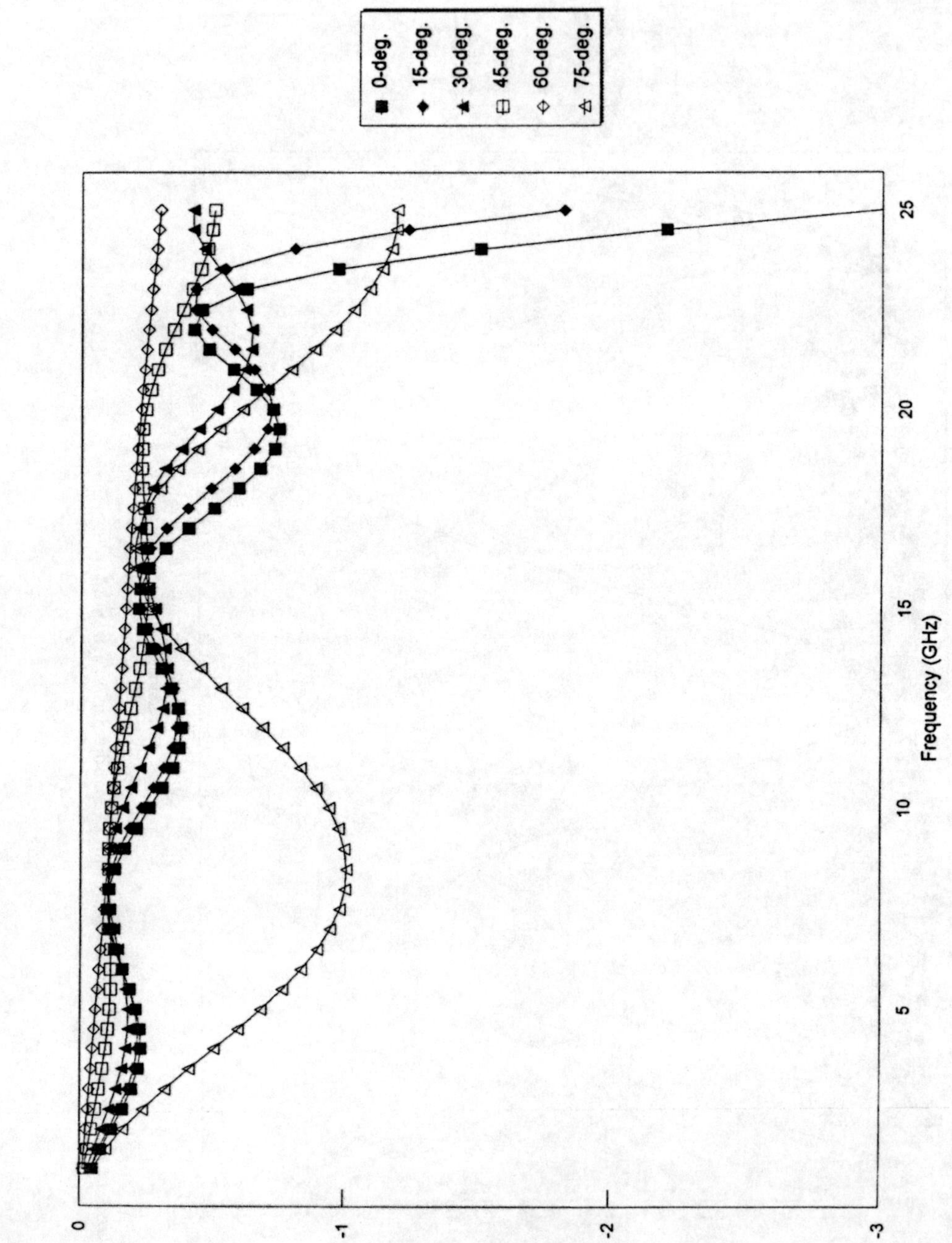

Figure 5.13 Seven-layer wall transmission with parallel polarization.

References

[1] Bodnar, D. G., and H. L. Bassett, "Analysis of an Isotropic Dielectric Radome," *IEEE Trans. Antennas Propagation,* Vol. AP-23, No. 6, November 1975.

[2] Yost, D. J., L. B. Weckesser, and R. C. Mallalieu, "Technology Survey of Radomes for Anti-Air Missiles," FS-80-022, John Hopkins Applied Physics Laboratory, Baltimore, MD, 1980.

[3] Collin, R. G., *Field Theory of Guided Waves,* New York: McGraw Hill Book Company, 1960.

[4] Kozakoff, D. J., "Reflectometer Measurement Approach for Radome Boresight Error and Loss Contours," *Proc. SPIE Int. Symp.,* Los Angeles, CA, 1994.

[5] Krueger, C. H., "A Computer Program for Determining the Reflection and Transmission Properties of Multilayer Plane Impedance Boundaries," AD 824 397, 1967.

[6] "Military Specification: Radomes, General Specification for," MIL-R-7705 B, January 12, 1975.

[7] Tornani, O., "Radomes, Advanced Design," North Atlantic Treaty Organization (NATO) Report AGARD-AR-53, prepared by Advisory Group for Aerospace Research and Development, Neuilly Sur Seine, France, 1973.

[8] Rudge, A. W., G. A. E. Crone, and J. Summers, "Radome Design and Performance," *Conf. Proc. Military Microwaves,* IEE (UK), TK6867.C65, 1980.

[9] Cary, R. H. J., "Design of Multi-Band Radomes," *Proc. IEE Conf. on Aerospace Antennas* (UK), IEE Publication No. 77, TL694.A6C57, 1971.

[10] Tice, T. E., "Techniques for Airborne Radome Design," Report AFATL-TR-66-391, U.S. Air Force, 1966.

Selected Bibliography

Walton, J. D., *Radome Engineering Handbook,* New York: Marcel Dekker, 1970.

APPENDIX 5A: PROGRAM WALL COMPUTER SOFTWARE LISTING

This program calculates transmission coefficient and the insertion phase delay for a multilayer dielectric wall.

Variable List

eo	= permittivity of free space
RAD	= conversion between radians and degrees
N	= number of layers
angle	= angle of incidence (degrees), measured from normal
f	= frequency (GHz)
finc	= frequency increment (GHz)
thk(N)	= thickness of the Nth layer (inches)
er(N)	= real part of relative permittivity of layer N
ltan(N)	= loss tangent of layer N
LAM	= wavelength (inches)
Ko	= free space wavenumber (inches ^ -1)
POL	= polarization flag
	= 0 for perpendicular (TE), 1 for parallel (TM)
Twdb	= transmission coefficient (dB)

```
ON ERROR GOTO error.in.program
start:
   REDIM Amag(10), Bmag(10), Cmag(10), Aph(10), Bph(10), Cph(10)
   REDIM zmag(10), zph(10), rez(10), imz(10), emag(10), eph(10), er(10),
 ltan(10)
   REDIM phimag(10), phipn(10), rephi(10), imphi(10)
   REDIM magterm(10), angterm(10), thk(10)
   REDIM RMAG(10), Rph(10), reR(10), imR(10)
   REDIM Tmag(10), Tph(10), reT(10), imT(10)
   REDIM Twdb(179, 90), IPD(179, 90)

   TYPE variables
      I AS INTEGER
      J AS INTEGER
      K AS INTEGER
   END TYPE

   eo = 8.854E-12
   PI = 3.14159265#
   RAD = PI / 180

'  Enter Data Here:
   COLOR 7, 0
   CLS
   LOCATE 1, 1, 1, 0, 15
   INPUT "Enter Computation Low Frequency (GHz)         : ", fl
   SOUND 600, .7
   INPUT "Enter Computation High Frequency (GHz)        : ", fh
   SOUND 600, .7
   INPUT "Enter Computation Frequency Increment (GHz) : ", finc
   SOUND 600, .7
   PRINT
   INPUT "Enter Number of Layers                        : ", N
   SOUND 600, .7

   FOR I = 1 TO N
     layer.msg$ = "For Layer Number " + LTRIM$(STR$(I)) + " "
     layer.msg$ = layer.msg$ + STRING$(80 -LEN(layer.msg$), 196)
     COLOR 15, 0
     PRINT
     PRINT layer.msg$
     COLOR 7, 0
     PRINT ""
     INPUT "  Enter Thickness (inches)                    : ", thk(I)
     SOUND 600, .7
     INPUT "  Enter Relative Permittivity                 : ", er(I)
     SOUND 600, .7
     INPUT "  Enter Loss Tangent                          : ", ltan(I)
     SOUND 600, .7
     IF CSRLIN >= 23 THEN CLS
```

```
    NEXT I
    COLOR 7, 0
    CLS

    INPUT "Do You Want to Send Data Output to a Printer              : ",
print.data$
    SOUND 600, .5
    CLS
    COLOR 15, 0
    CLS
    LOCATE 1, 1, 0

    PRINT ""
    PRINT " LAYER      THICKNESS     RELATIVE          LOSS"
    PRINT "  NO.       (INCHES)      PERMITIVITY       TANGENT"
    PRINT ""

      '—— print data to printer ——
      IF UCASE$(print.data$) = "Y" OR UCASE$(print.data$) = "YES" THEN
        LPRINT ""
        LPRINT " LAYER      THICKNESS     RELATIVE          LOSS"
        LPRINT "  NO.       (INCHES)      PERMITIVITY       TANGENT"
        LPRINT ""

        FOR I = 1 TO N
               LPRINT USING "  ##          #.###          ##.##
#.####"; I, thk(I), er(I), ltan(I)
        NEXT I
        LPRINT ""
      END IF
      '—— print data to printer ——
      FOR I = 1 TO N
        PRINT USING "  ##          #.###          ##.##           #.####";
I, thk(I), er(I), ltan(I)
      NEXT I
    COLOR 31, 0
    LOCATE 25, 25
    PRINT "- PRESS ENTER KEY TO CONTINUE -";
    DO UNTIL INKEY$ <> "": LOOP
    SOUND 800, .7
    COLOR 15, 0
    CLS

'-- Begin Calculations Here:
    FOR pol = 0 TO 1
       array.index = 0
      FOR f = fl TO fh STEP finc
        array.index = array.index + 1

        LAM = 11.811 / f
        Ko = .531976 * f
```

```
        FOR ANGLE = 0 TO 75 STEP 15
          theta = ANGLE * RAD
        GOSUB sub.wall
          Twdb(array.index, ANGLE) = 20 * LOG(Tw) / LOG(10)
          IPD(array.index, ANGLE) = -IPD / RAD
          IF IPD(array.index, ANGLE) = -360 THEN IPD(array.index,
 ANGLE) = 0
        NEXT ANGLE
      NEXT f

      IF pol = 0 THEN
        PRINT "Transmission Loss (in dB) for Perpendicular Polarization
Case"
        PRINT "(angle-of-incidence is a parameter)"
        PRINT ""
        PRINT " Freq    0-deg    15-deg   30-deg   45-deg   60-deg
75-deg"
        PRINT ""

          '-- printer output --
          IF UCASE$(print.data$) = "Y" OR UCASE$(print.data$) = "YES" THEN
            LPRINT ""
            LPRINT "Transmission Loss (in dB) for Perpendicular
Polarization Case"
            LPRINT "(angle-of-incidence is a parameter)"
            LPRINT ""
            LPRINT " Freq    0-deg    15-deg   30-deg   45-deg   60-deg
75-deg"
            LPRINT ""
          END IF
      ELSE
        PRINT "Transmission Loss (in dB) for Parallel Polarization Case"
        PRINT "(angle-of-incidence is a parameter)"
        PRINT ""
        PRINT " Freq    0-deg    15-deg   30-deg   45-deg   60-deg
75-deg"
        PRINT ""

          '-- printer output --
          IF UCASE$(print.data$) = "Y" OR UCASE$(print.data$) = "YES" THEN
            LPRINT ""
            LPRINT "Transmission Loss (in dB) for Parallel Polarization
Case"
            LPRINT "(angle-of-incidence is a parameter)"
            LPRINT ""
            LPRINT " Freq    0-deg    15-deg   30-deg   45-deg   60-deg
75-deg"
            LPRINT ""
          END IF
      END IF
```

```
        array.index = 0
        FOR f = fl TO fh STEP finc
        array.index = array.index + 1
'    Output Data to Screen:
           PRINT USING "  ##.#  +##.##   +##.##   +##.##   +##.##   +##.##
+##.##"; f, Twdb(array.index, 0), Twdb(array.index, 15),
Twdb(array.index, 30), Twdb(array.index, 45), Twdb(array.index, 60),
Twdb(array.index, 75)
'    Printer Output:
           IF UCASE$(print.data$) = "Y" OR UCASE$(print.data$) = "YES" THEN
             LPRINT USING "  ##.#  +##.##   +##.##   +##.##   +##.##
+##.##   +##.##"; f, Twdb(array.index, 0), Twdb(array.index, 15),
Twdb(array.index, 30), Twdb(array.index, 45), Twdb(array.index, 60),
Twdb(array.index, 75)
           END IF
         IF CSRLIN >= 24 THEN
           COLOR 31, 0
           LOCATE 25, 1
           PRINT STRING$(34, 196); "M o r e "; STRING$(35, 196);
           DO UNTIL INKEY$ <> "": LOOP
           SOUND 800, .7
           COLOR 15, 0
           CLS
         END IF

      NEXT f

      COLOR 31, 0
      LOCATE 25, 25
      PRINT "- PRESS ENTER KEY TO CONTINUE -";
      DO UNTIL INKEY$ <> "": LOOP
      SOUND 800, .7
      COLOR 15, 0
      CLS

      IF pol = 0 THEN
        PRINT "Insertion Phase Delay (in degrees) for Perpendicular
Polarization Case"
        PRINT "(angle-of-incidence is a parameter)"
        PRINT ""
        PRINT " Freq    0-deg   15-deg   30-deg   45-deg   60-deg
75-deg"
        PRINT ""

'    Printer Output:
           IF UCASE$(print.data$) = "Y" OR UCASE$(print.data$) = "YES" THEN
             LPRINT ""
             LPRINT "Insertion Phase Delay (in degrees) for Perpendicular
Polarization Case"
             LPRINT "(angle-of-incidence is a parameter)"
```

```
            LPRINT ""
            LPRINT " Freq    0-deg    15-deg   30-deg   45-deg   60-deg
75-deg"
            LPRINT ""
         END IF
      ELSE
         PRINT "Insertion Phase Delay (in degrees) for Parallel
Polarization Case"
         PRINT "(angle-of-incidence is a parameter)"
         PRINT ""
         PRINT " Freq    0-deg    15-deg   30-deg   45-deg   60-deg
75-deg"
         PRINT ""

'     Printer Output:
          IF UCASE$(print.data$) = "Y" OR UCASE$(print.data$) = "YES" THEN
            LPRINT ""
            LPRINT "Insertion Phase Delay (in degrees) for Parallel
Polarization Case"
            LPRINT "(angle-of-incidence is a parameter)"
            LPRINT ""
            LPRINT " Freq    0-deg    15-deg   30-deg   45-deg   60-deg
75-deg"
            LPRINT ""
         END IF
      END IF

        array.index = 0
      FOR f = fl TO fh STEP finc
        array.index = array.index + 1

'     Output Data to Screen:
          PRINT USING "  ##.#  +###.#   +###.#   +###.#   +###.#   +###.#
+###.#"; f,           IPD(array.index, 0), IPD(array.index, 15),
IPD(array.index, 30), IPD(array.index, 45), IPD(array.index, 60),
IPD(array.index, 75)

'     Printer Output:
          IF UCASE$(print.data$) = "Y" OR UCASE$(print.data$) = "YES" THEN
            LPRINT USING "  ##.#  +###.#   +###.#   +###.#   +###.#
+###.#   +###.#"; f, IPD(array.index, 0), IPD(array.index, 15),
IPD(array.index, 30), IPD(array.index, 45), IPD(array.index, 60),
IPD(array.index, 75)
          END IF

        IF CSRLIN >= 24 THEN
          COLOR 31, 0
          LOCATE 25, 1
          PRINT STRING$(34, 196); "<M o r e>"; STRING$(35, 196);
          DO UNTIL INKEY$ <> "": LOOP
          SOUND 800, .7
```

```
          COLOR 15, 0
          CLS
        END IF

      NEXT f

      COLOR 31, 0
      LOCATE 25, 25
      PRINT "- PRESS ENTER KEY TO CONTINUE -";
      DO UNTIL INKEY$ <> "": LOOP
      SOUND 800, .7
      COLOR 15, 0
      CLS

    NEXT pol

  '  Printer Output:
  IF UCASE$(print.data$) = "Y" OR UCASE$(print.data$) = "YES" THEN
    LPRINT CHR$(12)
  END IF
  LOCATE 1, 1, 0, 0, 0
  COLOR 11, 0
  CLS
 END

'    Wall Subroutine:
sub.wall:
    CTH = COS(theta)
    STH = SIN(theta) * SIN(theta)
    zmag(0) = 1
    zph(0) = 0
    rez(0) = 1
    imz(0) = 0
'    Calculate Complex Permitivity For Each Layer:
       FOR I = 1 TO N
        x = er(I)
        y = er(I) * -(ltan(I))
      GOSUB rect.to.polar
        emag(I) = MAG
        eph(I) = ANG
    NEXT I
    FOR I = 1 TO N
        x = er(I) - STH
        y = er(I) * (-ltan(I))
      GOSUB rect.to.polar
      GOSUB sub.complex.sqr.root
        magterm(I) = RMAG
        angterm(I) = RANG
        phimag(I) = Ko * magterm(I)
        phiph(I) = angterm(I)
        MAG = phimag(I)
```

```
      ANG = phiph(I)
     GOSUB sub.polar.to.rect
      rephi(I) = x
      imphi(I) = y
      zmag(I) = CTH / (magterm(I) + 1E-12)
      zph(I) = -angterm(I)
      MAG = zmag(I)
      ANG = zph(I)
     GOSUB sub.polar.to.rect
      rez(I) = x
      imz(I) = y
     IF pol = 1 THEN
       zmag(I) = 1 / ((emag(I) * zmag(I)) + 1E-12)
       zph(I) = -(eph(I) + zph(I))
       MAG = zmag(I)
       ANG = zph(I)
       GOSUB sub.polar.to.rect
       rez(I) = x
       imz(I) = y
    END IF
  NEXT I
   zmag(N + 1) = 1
   zph(N + 1) = 0
   rez(N + 1) = 1
   imz(N + 1) = 0
  FOR I = 1 TO (N + 1)
      x = rez(I) - rez(I - 1)
      y = imz(I) - imz(I - 1)
    GOSUB rect.to.polar
      NUMMAG = MAG
      NUMANG = ANG
      x = rez(I) + rez(I - 1)
      y = imz(I) + imz(I - 1)
    GOSUB rect.to.polar
      DENMAG = MAG
      DENANG = ANG
      RMAG(I) = NUMMAG / DENMAG
      Rph(I) = NUMANG - DENANG
      MAG = RMAG(I)
      ANG = Rph(I)
    GOSUB sub.polar.to.rect
      reR(I) = x
      imR(I) = y
      reT(I) = 1 + reR(I)
      imT(I) = imR(I)
      x = reT(I)
      y = imT(I)
    GOSUB rect.to.polar
      Tmag(I) = MAG
      Tph(I) = ANG
 NEXT I
```

```
 Amag(1) = EXP(-imphi(1) * thk(1))
 Amag(4) = 1 / Amag(1)
 Amag(2) = RMAG(1) * Amag(4)
 Amag(3) = RMAG(1) * Amag(1)
 Aph(1) = rephi(1) * thk(1)
 Aph(2) = Rph(1) - Aph(1)
 Aph(3) = Rph(1) + Aph(1)
 Aph(4) = -Aph(1)
FOR K = 2 TO N
    Bmag(1) = EXP(-imphi(K) * thk(K))
    Bmag(4) = 1 / Bmag(1)
    Bmag(2) = RMAG(K) * Bmag(4)
    Bmag(3) = RMAG(K) * Bmag(1)
    Bph(1) = rephi(K) * thk(K)
    Bph(2) = Rph(K) - Bph(1)
    Bph(3) = Rph(K) + Bph(1)
    Bph(4) = -Bph(1)
  GOSUB sub.cmult
NEXT K
 Bmag(1) = 1
 Bmag(4) = 1
 Bmag(2) = RMAG(N + 1)
 Bmag(3) = Bmag(2)
 Bph(1) = 0
 Bph(2) = Rph(N + 1)
 Bph(3) = Bph(2)
 Bph(4) = 0
GOSUB sub.cmult
 tranmag = 1
 tranph = 0
FOR J = 1 TO N + 1
   tranmag = tranmag * Tmag(J)
   tranph = tranph + Tph(J)
NEXT J
 tranmag = 1 / (tranmag + 1E-12)
DO
  IF tranph > (2 * PI) THEN
    tranph = tranph - (2 * PI)
  ELSE
    EXIT DO
  END IF
LOOP
 tranph = -tranph
 Tw = 1 / (tranmag * Amag(1))
 Ttotph = -(tranph + Aph(1))
 SUM = 0
FOR J = 1 TO N
   SUM = SUM + thk(J)
NEXT J
'-- Get IPD into Correct Quadrant
```

```
     DO
        IF Aph(1) < (CTH * Ko * SUM) THEN Aph(1) = Aph(1) + (2 * PI)
        IF Aph(1) >= (CTH * Ko * SUM) THEN EXIT DO
     LOOP
      IPD = Aph(1) - CTH * Ko * SUM
     RETURN

'-- Cmult Subroutine Starts Here:
sub.cmult:
        FOR I = 1 TO 3 STEP 2
       FOR J = 1 TO 2
         R1 = Amag(I) * Bmag(J)
         R2 = Amag(I + 1) * Bmag(J + 2)
         E1 = Aph(I) + Bph(J)
         E2 = Aph(I + 1) + Bph(J + 2)
         R3 = R1 * COS(E1) + R2 * COS(E2)
         E3 = R1 * SIN(E1) + R2 * SIN(E2)
         Cmag(J + I - 1) = SQR(R3 * R3 + E3 * E3)
         IF (R3 > 0) AND (E3 > 0) THEN PA = 0
         IF (R3 < 0) THEN PA = PI
         IF (R3 > 0) AND (E3 < 0) THEN PA = 2 * PI
         Cph(J + I - 1) = (ATN(E3 / (R3 + 1E-12))) + PA
       NEXT J
     NEXT I
     FOR I = 1 TO 4
        Amag(I) = Cmag(I)
        Aph(I) = Cph(I)
     NEXT I
     RETURN

'-- Rectangular to Polar Subroutine Starts Here:
rect.to.polar:

      MAG = SQR(x ^ 2 + y ^ 2)
     IF (x > 0) AND (y > 0) THEN corr = 0
     IF (x < 0) AND (y < 0) THEN corr = PI
     IF (x > 0) AND (y < 0) THEN corr = 0
     IF (x < 0) AND (y > 0) THEN corr = PI
     IF (x > 0) AND (y = 0) THEN corr = 0
     IF (x < 0) AND (y = 0) THEN corr = PI
     IF (x = 0) AND (y <> 0) THEN corr = 0
     IF (x = 0) AND (y = 0) THEN
        ANG = 0
        GOTO exit.rect.to.polar
     END IF
      ANG = ATN(y / (x + 1E-12)) + corr
exit.rect.to.polar:
     RETURN

'-- Polar to Rectangular Subroutine Starts Here:
sub.polar.to.rect:
```

```
     x = MAG * COS(ANG)
     y = MAG * SIN(ANG)
    RETURN

'-- Complex Square Root Subroutine Starts Here:
sub.complex.sqr.root:
     RMAG = SQR(MAG)
     RANG = ANG / 2
    RETURN

error.in.program:
  COLOR 31, 0
  CLS
  LOCATE 13, 10
  PRINT "ERROR  -  Input Variables Outside Constraints of Program..."
  PLAY "L10 N29 N25 N29 N25 N29 N25"
  DO UNTIL INKEY$ <> "": LOOP
  COLOR 11, 0
  CLS
  END
```

Chapter 6

Radome Analysis Techniques

This chapter summarizes selected radome analysis techniques, applied to prediction of performance for radome-enclosed antennas. In any analysis technique, the most significant electrical performance parameters of a radome are (1) boresight error, (2) copolarized and cross-polarized transmission loss, and (3) antenna sidelobe-level degradation. For certain applications, other parameters that might be of concern are antenna pattern ellipticity ratio and antenna voltage standing wave ratio (VSWR) [1].

Ray tracing is a widely used method for describing propagation through a radome wall. When using ray tracing in the receiving mode, an incident wavefront is sampled at uniform sample increments, generally a half-wave apart. From each of these sample points, a ray in the direction of the propagation vector is traced through the radome wall and onto the antenna aperture where integration of these rays produces the antenna receiving port voltage. The point at which each of these rays intercepts the outer surface of the radome is known as the *intercept point.*

Ray-tracing methods use many approximations, such as the idea of treating the radome wall at each intercept point as locally plane, and assuming parallelism between the inner and outer radome walls at the intercept point. Several other factors, such as the antenna size and radome curvature (in wavelengths), may influence discrepancies between the measured and predicted results. Because of the many other approximations used in ray tracing, the results have limited validity that can only be assessed by comparing measured and predicted results.

6.1 BACKGROUND

Early methods of radome electrical design were cumbersome and approximate, relying primarily on the use of nomographs, such as Kaplun [2] and Zamyatin et al. [3]. With the arrival of the microcomputer, computer-aided design of radomes eventually became widespread [4]. The earliest recorded use of a digital computer in radome analysis is believed to be a 1959 paper by Mahan et al. [5]. Other early design methods that used computers were described in papers by Tricoles [6,7].

Although a number of different radome electromagnetic analysis techniques are feasible, this book focuses primarily on the geometric optics (GO) and the physical optics (PO) methods that can be easily implemented on a personal computer (PC). GO treats the electromagnetic propagation as being "light like" in behavior and, although approximate, is exact only in the limit of zero wavelength (infinite frequency.) However, it produces reasonably good results for radome-enclosed antennas as small as about five wavelengths in diameter. GO will not give good results near physical boundaries, but provides useful insights in those cases. We must consider three aspects to fully use GO: (1) ray reflections, (2) polarization, and (3) amplitude variations along the ray path and through reflections [8].

PO is based on Huygens's principle, which has been of fundamental importance to the development of electromagnetic wave theory. Huygens's principle states that each point on a primary wavefront can be considered as a new source of secondary spherical waves, and that a secondary wavefront can be constructed as the envelope of these spherical waves. Thus, a spherical wave from a point source propagates as a spherical wave as illustrated in Figure 6.1(a), whereas an infinite plane wave continues as a plane wave as depicted in Figure 6.1(b). PO methods are employed in surface integration formulations that produce reasonably good results for radome-enclosed antennas less than a wavelength in diameter.

A comparison between PO and GO solutions can be obtained by considering a uniform plane electromagnetic wave incident on the electrically large and perfectly conducting obstacle shown in Figure 6.2(a). We want to obtain a solution to the electric field variation at points along the x axis and at a distance r behind the obstacle. According to Kraus [9], we can apply Huygens's principle to obtain:

$$E^i = \int_{\text{along x-axis}} dE \tag{6.1}$$

where dE is the electric field at point P due to a point source at distance x from the origin as illustrated in the figure. We note that in terms of wave number k, the following holds true:

$$dE = \frac{E_o}{r} e^{-jk(r+\delta)}\, dx \tag{6.2}$$

so that

$$E^i = \frac{E_o}{r} e^{-jkr} \int_{-\infty}^{\infty} e^{-jk\delta}\, dx \tag{6.3}$$

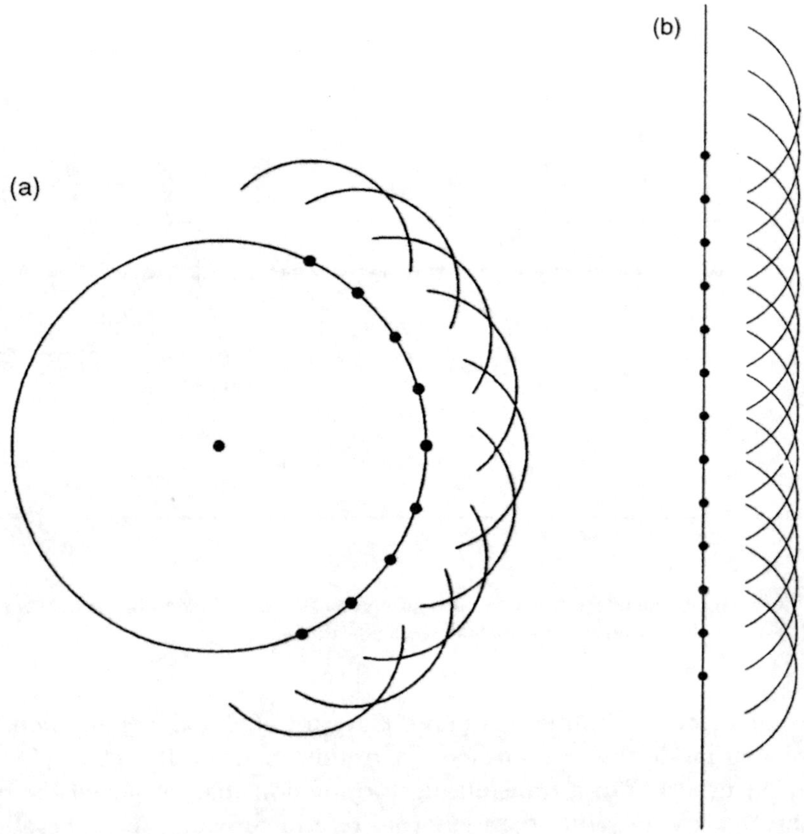

Figure 6.1 Illustration of Huygens's sources: (a) point source and (b) plane wave.

The PO solution for the power density variation at a distance *r* below the plane is shown in Figure 6.2(b). Also depicted is the GO solution wherein it is seen that the conducting obstacle casts a simple shadow, illustrating its light-like behavior. Intuitively, the PO approach is more robust. For instance, assume an incident wavefront on a radome having a conducting rain erosion tip. The GO modeling approach would cast a simple tip shadow onto the antenna aperture. The PO modeling approach is more robust in predicting the actual electromagnetic (EM) fields incident on the antenna aperture.

Section 6.2 discusses the GO method, applied by Kilcoyne to develop a two-dimensional ray-tracing code for analyzing radomes on a digital computer [10]. Bagby subsequently published a three-dimensional GO method applicable to desktop computer-aided design of aircraft radomes [11].

Section 6.3 describes PO methods, which were developed later and offer more

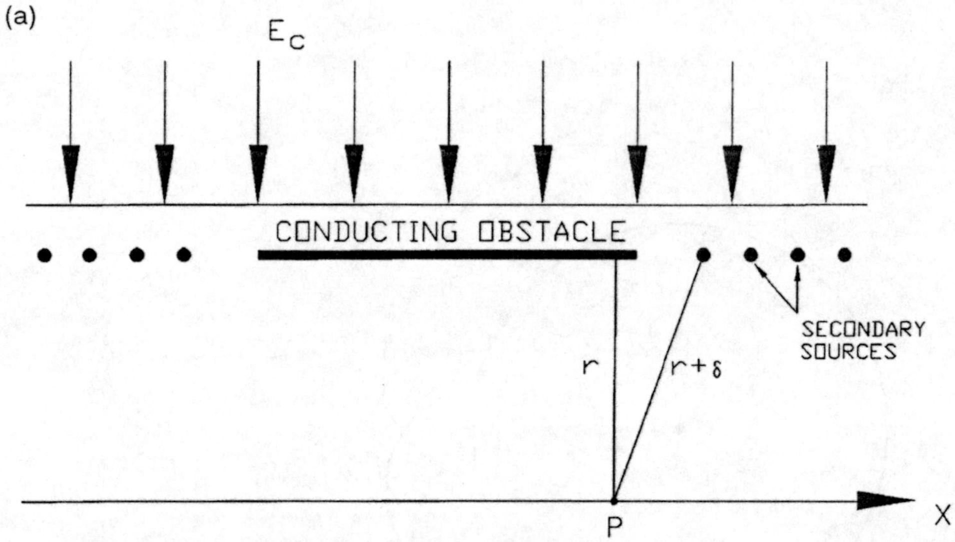

Figure 6.2 Plane wave incident from above onto an electrically large plane object: (a) geometry and (b) power density variation at distance or below the plane.

robust radome electromagnetic modeling. Hayward et al. [12,13] adapted PO to a simple receiving mode radome surface integration method. Raz et al. [14] and Israel et al. [15] used PO in a transmitting formulation that projected the antenna aperture through the radome to an external reference plane. An integration over this reference plane produced the far-field radiation patterns of the radome-enclosed antenna.

Section 6.4 describes other methods that can be employed for radome analysis, for instance, the method of moments (MOM) technique reported by Tricoles et al. [16] and the plane wave spectra (PWS) approach used by Wu and Rudduck [17] and Joy and Huddleston [18].

Section 6.5 discuses sources of computational errors in the modeling. Primarily, these include those error sources pertinent to the GO and PO analysis approaches. A comparison of the relative accuracy and difficulty of implementation of the techniques discussed in this chapter appears in [12,13,19,20].

6.2 GEOMETRIC OPTICS APPROACHES TO PROPAGATION THROUGH RADOMES

Most radome analysis approaches based on GO ray tracing have the following characteristics: (1) conceptual simplicity combined with reasonable accuracy and (2) application in either the transmitting or receiving mode.

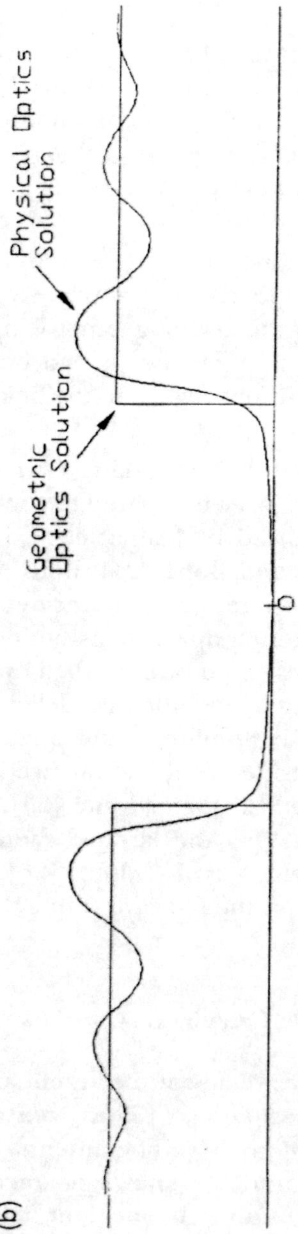

Figure 6.2 (continued)

For radomes large in wavelengths, GO methods produce good boresight error prediction accuracy. Tavis [21] researched both the transmitting and the receiving GO formulations and found identical boresight error prediction with either model. However, the receiving approach required greater computer run time.

According to GO, we can decompose the problem of defining the EM field with a radome over the antenna into two parts:

1. Determine the rays that define the phase of the field independently from its amplitude. The ray optic field is a local plane-wave field that propagates along trajectories satisfying the laws of geometric optics [22].
2. Determine the amplitude that can be carried out by following intensity variations along each ray without regard to the field of other rays [23].

The earliest known approach to radome analysis was an optical one for which the aperture was assumed to be very large compared to a wavelength. A solution for a two-dimensional wedge was reported by Mahan et al. [5] and a solution for a three-dimensional cone by Mahan and Bone [24]. Both considered a monolithic radome wall, and the approach traced rays from an incident wavefront through the radome walls onto an internal antenna aperture assuming that the refraction effects at the air/dielectric boundaries could be described by Snell's law.

Another early GO transmitting formulation modeled the antenna aperture as a collection of sources with arbitrary amplitude and phase. The collection of rays defined an equivalent aperture outside of the radome whose amplitude and phase distributions incorporated the effect of the radome [25]. The fast Fourier transform (FFT) of this distribution yielded the far-field radiation pattern of the radome-enclosed antenna. In Einziger and Felsen [26], GO ray tracing also accounted for multiple reflections on the concave (inner) radome side, and for reflections between its layer boundaries.

Geometrical Optics Receiving Mode Calculations

The receiving formulation finds the voltage at the antenna port when a plane wave is incident on a radome-enclosed antenna, as illustrated in Figure 6.3(a). This is sometimes referred to as a *backward ray trace.* The antenna receiving port voltage is obtained from the incident fields and the known characteristics of the antenna. Thus, the problem is really one of finding the incident fields at the antenna in the presence of both the radome and the antenna [27]. For monopulse radar antennas, we can derive a quantitative estimate of the radome boresight error from the complex sum and difference antenna patterns computed without and with the radome in place.

Assume that an incident plane EM wave is coming from an arbitrary but fixed direction, θ, ϕ. To obtain the received antenna voltage, we integrate the wavefront data over the receiving antenna aperture in accordance with the equivalence theorem [6].

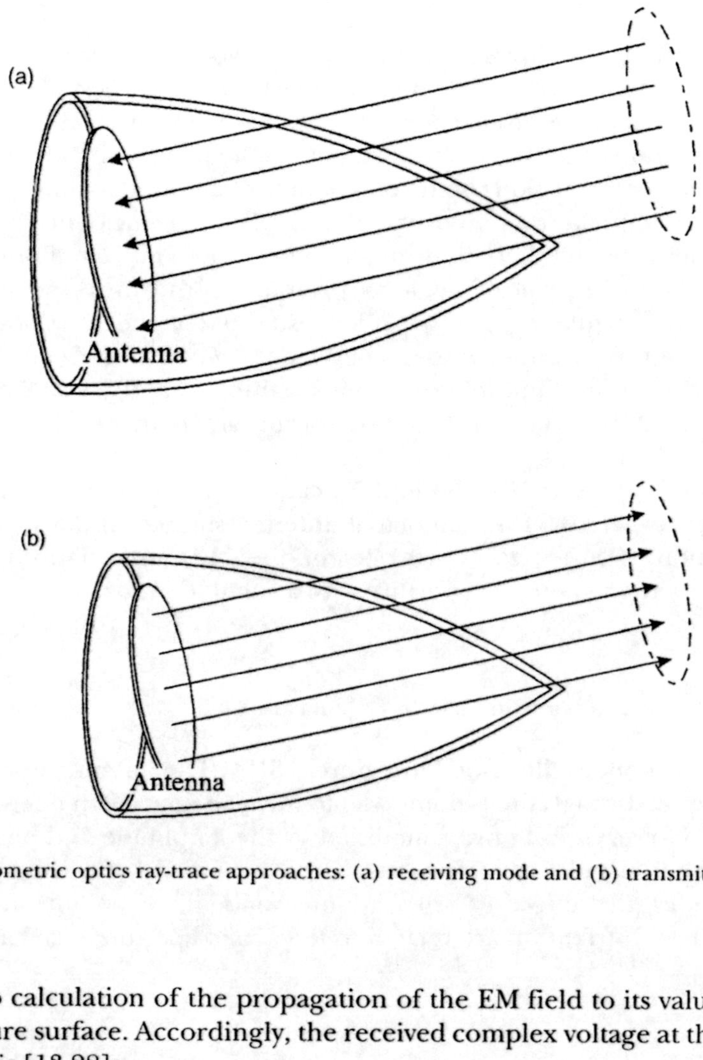

Figure 6.3 Geometric optics ray-trace approaches: (a) receiving mode and (b) transmitting mode.

This leads to calculation of the propagation of the EM field to its values on the antenna aperture surface. Accordingly, the received complex voltage at the terminal of the antenna is [13,28]:

$$V_a = \iint F^a (E^i T_W)\, da \tag{6.4}$$

This integration extends over the physical antenna surface area for which F^a has a significant contribution. In addition, E^i is the incident plane wave function; F^a is the complex valued receiving aperture distribution at point on the antenna surface; and T_W is the complex valued radome transmission coefficient of a ray intercept point.

Usually, the radome is considered locally plane at each ray intercept point [29], which leads to small errors in the calculation of T_W. The transmission characteristics of the radome wall for any given frequency, material type, number of layers, and layer thicknesses also depend on the angle-of-incidence θ (the angle between the propagation vector and the radome surface normal vector at the ray intercept point) and the apparent polarization of the wave E^i with respect to the plane of incidence. The term $T_{W\|}$ is the complex transmission coefficient for field components along the incidence plane, and $T_{W\perp}$, is for field components perpendicular to the incidence plane. We will see in a subsequent chapter that E^i is normally decomposed into components parallel and perpendicular to the plane of incidence prior to applying the respective transmission coefficients. Subsequently, the wave is reconstructed on the other side of the wall. For all cases, an equivalent transmission coefficient T_W can be defined. The calculation of T_W usually ignores multiple reflections within the radome wall and trapped waves. Tricoles et al. [30] have observed significant errors that occur sometimes by this omission.

The antenna aperture distribution, F^a, can be found from probing the near field at several wavelengths from an actual antenna surface, and applying a backward propagation (Fourier transform) technique [12]. Alternatively, it can be probed very close to the antenna aperture with a suitable probe.

Geometrical Optics Transmitting Mode Calculations

The transmitting mode is illustrated in Figure 6.3(b). The antenna aperture distribution is projected through the radome wall to form an equivalent aperture outside of the radome. However, each ray is modified by the amplitude and phase of its associated radome transmission coefficients, and the modified effective aperture distribution includes the effect of the radome walls. Thus, points forming the modified aperture distribution are related to the actual aperture distribution via:

$$F_{mn}^{a'} = T_W F_{mn}^{a} \tag{6.5}$$

where, F_{mn}^{a} is the complex valued transmit aperture distribution at a point on the actual antenna aperture, and T_W is the complex valued radome transmission coefficient associated with the intercept of the mn'th ray at its intercept point. In this case, the equivalent aperture size is identical in size to the actual antenna aperture. Assume that the antenna is in the x-y plane with the z-coordinate corresponding to the radome axis. We can obtain the far-field antenna array pattern from the reference plane distribution via:

$$E_t = \sum_{m=1}^{M} \sum_{n=1}^{N} F_{mn}^{a'} e^{-jk(x_m \sin\theta\cos\phi + y_n \sin\theta\sin\phi)} \tag{6.6}$$

where $x_m = m\, d_x$; $y_n = n\, d_y$; and d_x, d_y are the sample spacings in the x and y coordinate directions, respectively.

6.3 PHYSICAL OPTICS APPROACHES TO PROPAGATION THROUGH RADOME WALLS

In analyzing the performance of a radome-enclosed antenna system, PO techniques generally offer a higher computational accuracy than GO because of the necessary assumptions used with GO. For instance, GO makes the assumption that the EM wave propagates as a plane wave confined to a cylinder whose cross section the antenna aperture defines. However, in reality the field diverges.

Physical Optics Receiving Mode Calculations

For radomes smaller than about five wavelengths in diameter, the GO approach is too approximate and we must go to PO techniques. Integration over the radome surface with the Kirchoff–Fresnel integral yields better results to obtain the field at each point on the antenna aperture, as illustrated in Figure 6.4(a). To accomplish this integration, we use an external reference plane to reformulate an incident plane wave as a grid of Huygens's sources, and we then trace rays between that grid and each point on the antenna aperture. Hence,

$$V_a = \iint F^a \left(\iint E^i T_W \frac{e^{-jkr}}{r}\, ds \right) da \tag{6.7}$$

The outer integration extends over the antenna surface area for which F^a has significant contribution, and the inner integration is over the external reference plane where:

- Ei = Incident plane wave function;
- F^a = Complex valued receiving aperture distribution at a point on the antenna surface;
- T_W = Complex valued radome transmission coefficient at a ray intercept point;
- k = Wave number = $2\pi/\lambda$;
- r = The distance from each point on the external reference plane to an antenna aperture point, x_a, y_a, z_a.

The direct ray (GO) and surface integration (PO) methods differ primarily in the computation of the total transmission coefficient. With the PO method, the fact

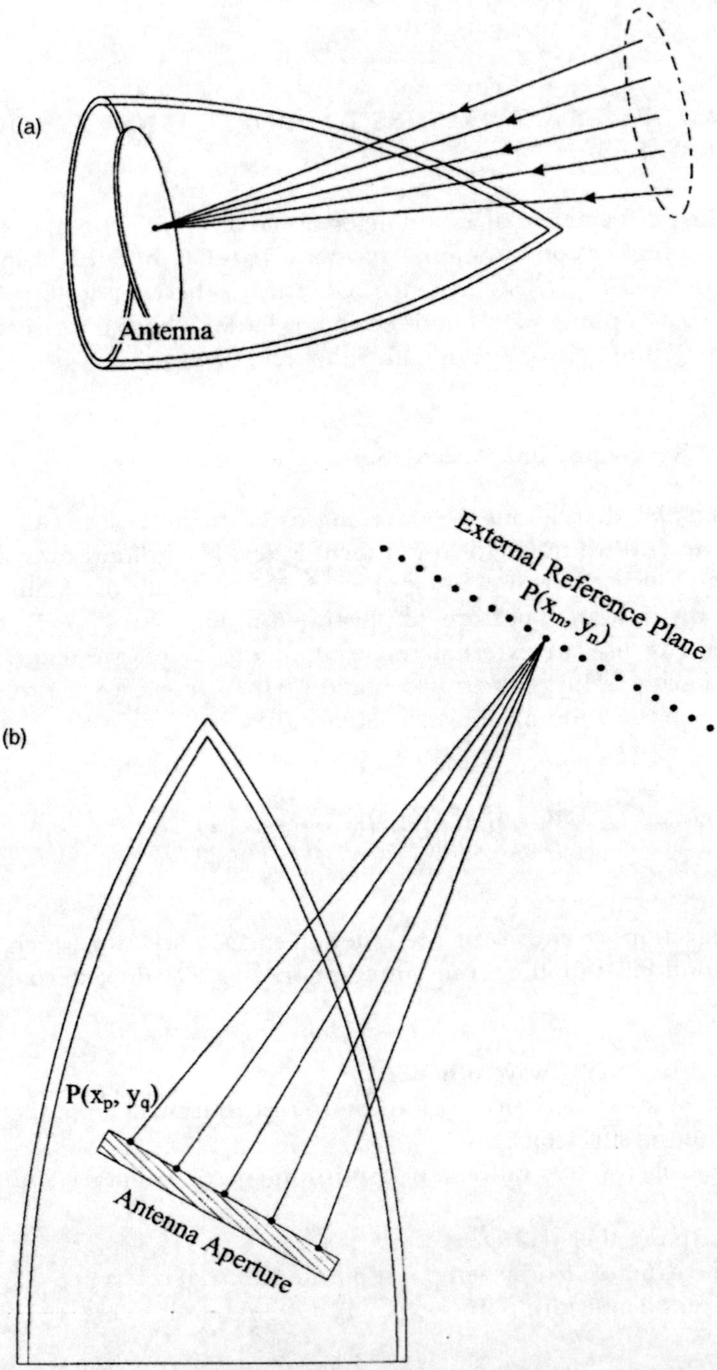

Figure 6.4 Physical optics: (a) receiving mode and (b) transmitting mode.

that the integration of a bundle of rays through the radome wall, as opposed to a single ray, more densely samples the curvature variation results in a more robust wall transmission model.

Both methods use the flat-panel approximation to compute wall transmission coefficients at the ray intercept points and ignore both multiple internal reflections and trapped waves. Hayward et al. [12] compared predicted with measured data for these two receiving ray-trace methods. Overall, the researchers found that the surface integration was more accurate, especially for radomes that were small in wavelengths.

Physical Optics Transmitting Mode Calculations

In a PO transmitting formulation, the following steps occur, which conform better than GO to reality:

1. We obtain the antenna near-field distribution on the inner radome surface.
2. The transmission coefficients of the radome wall are applied to give the field on the exterior of the radome.
3. We make a final transformation to the far field using an FFT technique.

A useful radome analysis technique researched by Raz et al. [14] and Israel et al. [15] uses PO for tracing the field from an antenna aperture through the radome wall to an external reference plane, where contributions are collected. The geometry is illustrated in Figure 6.4(b).

Assuming that each point on the antenna aperture radiates as a Huygens's source, the complex distribution can be derived for each point on an external reference plane via:

$$F^{a'}_{mn} = \iint F^a T_W \frac{e^{-jkr}}{r}\, da \qquad (6.8)$$

The integration extends over the physical antenna surface area for which F^a provides a significant contribution, and where

$F^{a'}_{mn}$ = Complex valued transmitting aperture distribution at a point on the external reference plane surface;
T_W = Complex valued radome transmission coefficient at ray intercept point;
k = Wave number = $2\pi/\lambda$;
r = The distance from each point on the antenna aperture to a point x_m, y_n on the external reference plane.

Here, the area forming the external reference plane should be larger than the physical area of the radome-enclosed antenna. An exact solution corresponds to an infinite external plane, but this is not possible. Typically, we may take it two to four times the physical area of the antenna. The far-field antenna array pattern can be computed from the reference plane distribution as we did in (6.6):

$$E_t = \sum_{m=1}^{M} \sum_{n=1}^{N} F_{mn}^{a'} e^{-jk(x_m \sin\theta \cos\phi + y_n \sin\theta \sin\phi)} \tag{6.9}$$

where $x_m = m\ d_x$; $y_n = n\ d_y$; and d_x, d_y are the sample spacings in x and y coordinate directions, respectively.

6.4 OTHER TECHNIQUES

Method of Moments

Joy et al. [20] applied a two-dimensional method of moments (MOM) technique to a tangent ogive radome, based on the approach developed by Richmond [27–29]. An integral equation for the electric field was solved by dividing the dielectric into cells that were small enough so that the electric field intensity was approximately uniform in each cell.

Also applying a MOM approach, Tricoles et al. [30] analyzed the EM fields scattered by a hollow cone, using two formulations. One was based on a scalar Green's function and the other on a tensor Green's function. The two are equivalent, but the procedures differ in the following ways:

- The scalar Green's function method decomposes the cone into circular cylinders, and subsequently decomposes the cylinders into angular sectors. The tensor Green's function method decomposes the cone into spheres.
- The number of cells differs because the cell sizes differ.
- Polarization dependance differs.

More recently, Tricoles et al. [16] applied the MOM technique to dielectric cylinders of arbitrary shape.

Although sophisticated, the MOM is more difficult to apply to radome modeling than the other methods previously discussed. Very limited validation data appear in the technical literature to compare its accuracy with the geometric optics of physical optics methods. However, the method of moments may give the effects of guided waves or scattering by a radome rain erosion tip.

Plane Wave Spectra

Paris developed a three-dimensional approach that could find the tangential fields on the outside radome surface because of a horn antenna radiating inside the

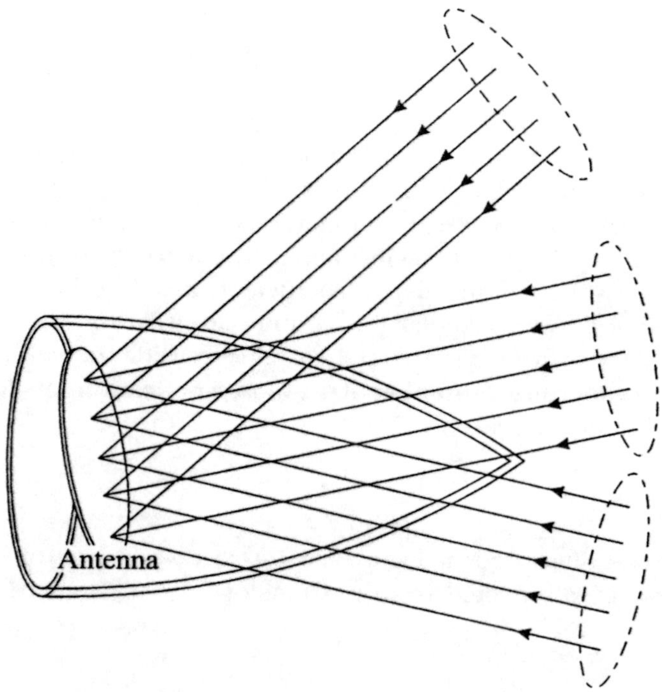

Figure 6.5 Illustration of plane wave spectra approach.

radome [31]. Wu and Rudduck developed a three-dimensional method that used a plane wave spectrum representation to characterize the antenna [17]. A discrete plane wave spectrum (PWS) is a complex vector array obtained from a Fourier transform of the near-field aperture and represents the radiating antenna properties. If an antenna aperture has $M \times N$ sample points, then the PWS contains MN plane waves. The PWS approach is illustrated in Figure 6.5.

Joy and Huddleston expanded on this idea by using a FFT to speed the computer calculations when applying a plane wave spectrum approach [18]. PWS is believed to give more accurate results than either GO or PO approaches discussed earlier. However, PWS requires considerably larger amounts of computer run time and cannot be implemented on a PC.

6.5 SOURCES OF RAY-TRACE COMPUTATIONAL ERROR

Ray-trace computational errors have several causes, including the following:

- Antenna modeling;
- Statistical wall variations;

- Internal wave reflections;
- Rain erosion tip modeling;
- Radome wall modeling.

Table 6.1 gives an assessment of additional factors contributing to radome performance computational accuracy. The internal reflections illustrated in Figure 6.6 are a potentially significant source of computational error. In this figure, a GO receiving mode ray trace uses the direct rays incident on the aperture from a specified direction. The insertion voltage transmission coefficients based on flat-panel theory for both the parallel and perpendicular polarization are used to transform the plane wave fields associated with each ray from their values outside the radome to their values on the antenna aperture. A number of the most significant errors (above) are discussed in the following paragraphs.

Internal Reflections

Sidewall reflections illustrated in Figure 6.6(b) exist when part of the energy incident on a radome is reflected from portions of the wall into the antenna. For

Table 6.1
Parameters That Influence Prediction Accuracy

Category	*Uncertainties*
Antenna modeling	Antenna aperture distribution
	Polarization detail
	Physical detail
Statistical wall variations	Thickness variations
	Dielectric constant Inhomogeneity
	Loss tangent uncertainty
Internal reflections	Back wall reflections
	Bulkhead reflections
	Multiple bounce
	Flashlobe
Tip modeling	Refraction
Wall model	Flat-panel assumption
	Anisotropic materials
	Variation of permittivity with temperature
	Adhesion layers
	Trapped waves

(a)

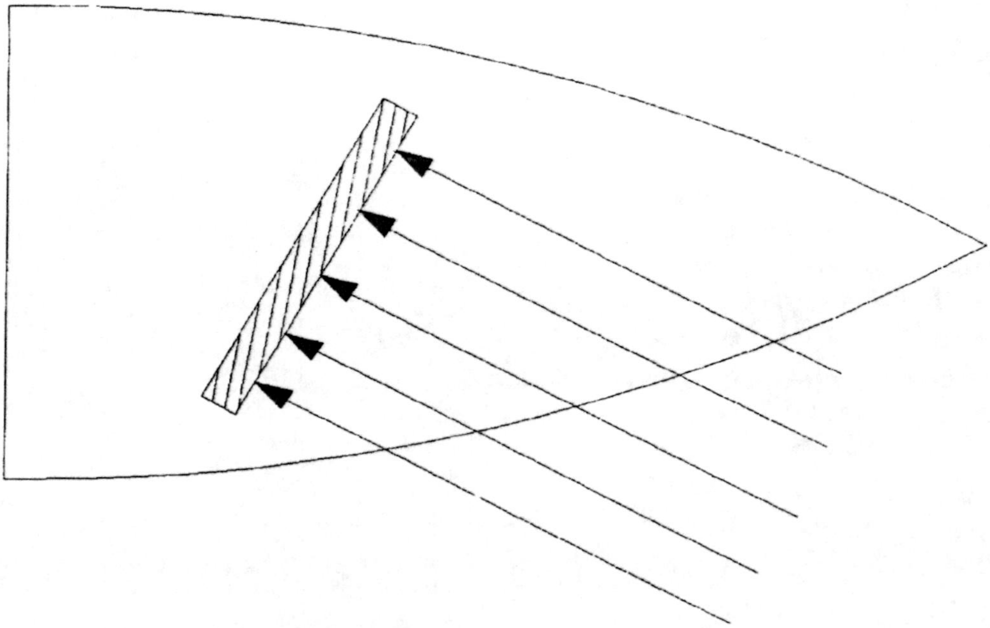

(b)

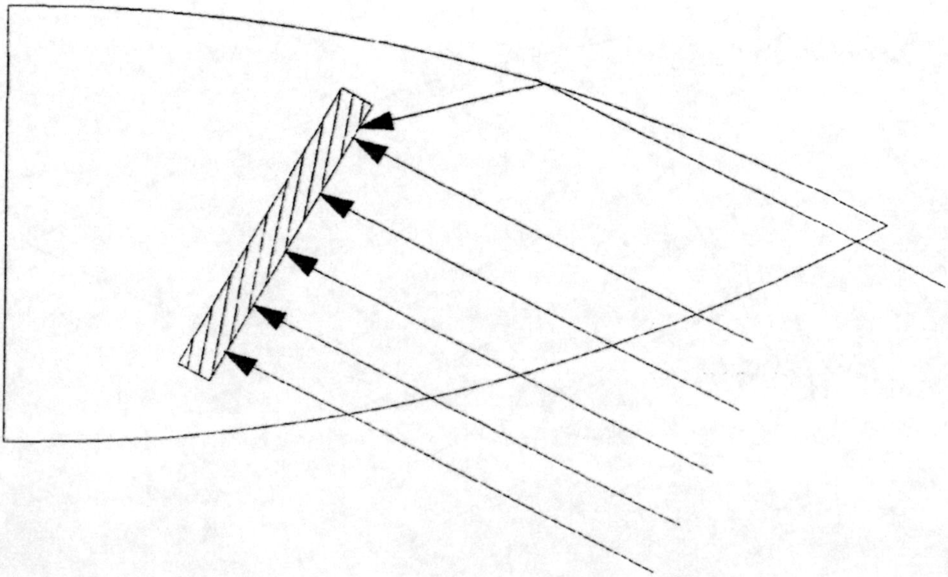

Figure 6.6 Internal ray reflections that deteriorate computational accuracy: (a) backward ray trace, (b) sidewall reflections, (c) antenna scattered energy, and (d) bulkhead reflections.

(c)

(d)

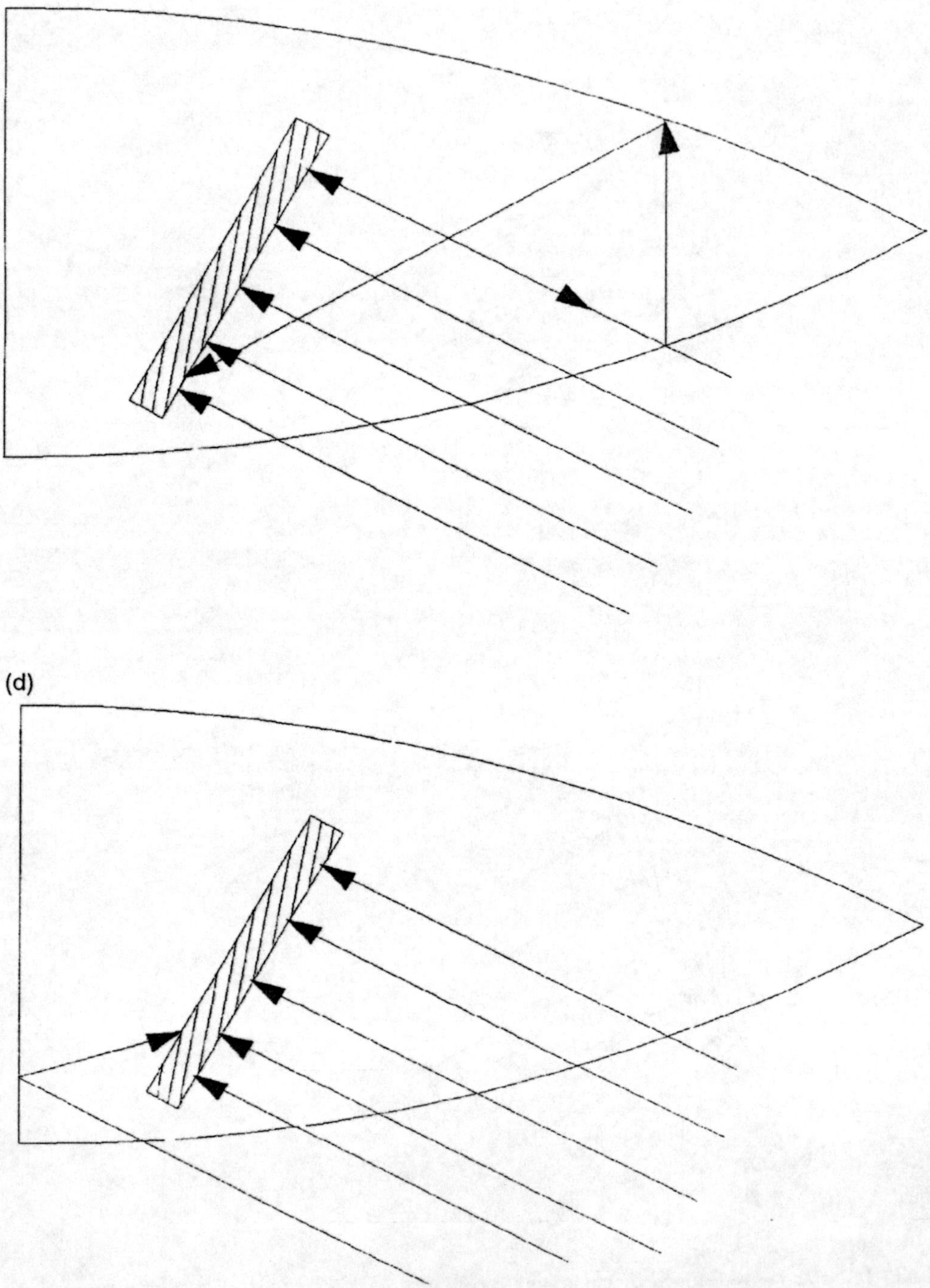

Figure 6.6 (continued)

most radomes, it is safe to ignore rays that reflect off the radome wall more than once before striking the antenna because these multiple bounces produce insignificant energy. The mathematical modeling describing this error source is shown in Burks et al. [32].

Figure 6.6(c) illustrates antenna scattered energy. Notice that not all the energy striking an antenna is absorbed by the antenna. The amount absorbed depends on the antenna's aperture distribution. The inverse of the antenna's illumination function often provides an approximation to the energy reflected by the antenna aperture. The antenna scattered energy can bounce off the radome wall one or more times before striking the antenna. Again, rays that reflect off the radome wall more than once before striking the antenna are relatively small and can be ignored.

Many aircraft or missiles have a bulkhead plate between the radome and the vehicle body, as illustrated in Figure 6.6(d). If this bulkhead is not microwave-absorber treated, incoming rays can reflect off this surface and strike the backside of the antenna. The backlobe pattern of the antenna receives the , causing boresight errors. Generally, bulkhead reflections will not have much effect until the angle-of-incidence of the incoming rays (with respect to the radome axis) becomes large. In actual applications, positioners and other electronic hardware may block some reflections from reaching the antenna.

To evaluate the typical contribution of each of the ray bounce error contributors, we studied an A-sandwich tangent ogive radome with a fineness ratio of three. The antenna size was approximately 20 wavelengths at the operating frequency and the polarization was vertical. Using a GO receiving mode prediction model, we compared the measured elevation plane BSE data to predicted data in Figure 6.7. For this

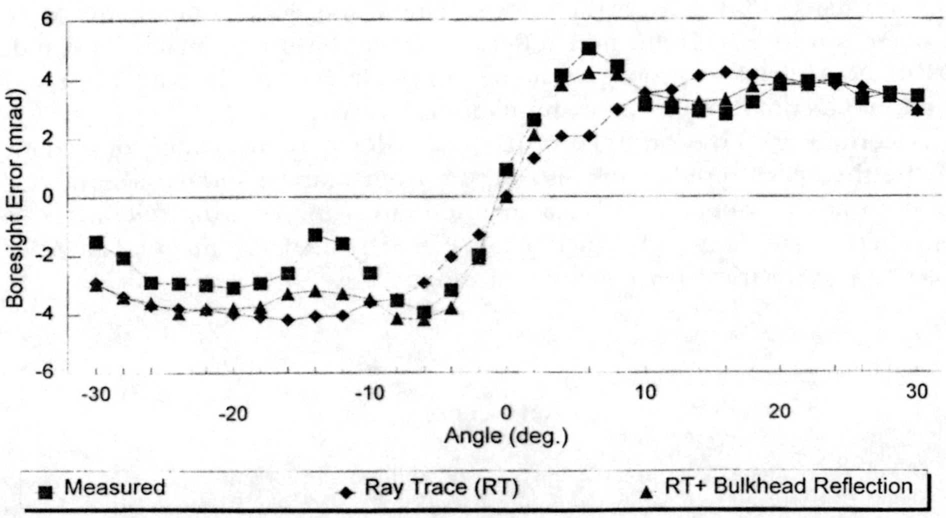

Figure 6.7 Measured and computed BSE of an A-sandwich tangent ogive radome and the effect of bulkhead reflections.

particular radome, when ignoring all internal reflection error sources, reasonably good correlation was seen between measured and predicted BSE data. The predominant internal reflection error source was found to be bulkhead reflections. The computed BSE data when including this error source are also shown in the figure.]

Wall Model and Statistical Variations

Another source of computational error is the wall transmission model. In all current ray-trace solutions, the radome walls are approximated as locally plane at the ray intercept point because most radome shapes are not separable coordinate surfaces, and differential equations cannot be used. The flat-panel approximation is not valid for thick, highly curved walls. For spherical radomes, Bloom et al. [33], attempted to compensate the transmission coefficient to account for wall curvature by use of a divergence factor (DF). Kozakoff [34] derived a solution applicable to ogive radome shapes. Geometric optics was used to trace the fields from the reference plane through the radome wall to a receiving monopulse antenna, where the wall transmissions on each ray collected. However, a DF derived from Snell's law for spherical shells accounted for the local wall curvature at each ray intercept point.

Most radome wall models assume that the dielectric and magnetic properties are homogeneous and isotropic with individual layers. This is not always true, and is another potential source of computational error and complication in the modeling. For instance, Kozakoff and Hensel [35] showed that the orientation of structural fibers in reinforced plastic radomes results in anisotropy of the electromagnetic medium that influences the modeling. To solve this type of problem, White and Banks [36] derived the wave equation into three components to deal with plane wave transmission and reflection from anisotropic materials. Guided waves were observed to propagate in radome walls by Tricoles and Rope [37], and are a potential source of computational error.

According to Ersoy and Ford [38], a radome performance degradation from the theoretical predictions also occurs from radome surface irregularities. Vorobyev [39] has applied statistical methods to manufacturing tolerances (on both dielectric constant and wall thickness) and related the manufacturing tolerances to the electrical performance of radomes.

References

[1] Torani, O., "Radomes, Advanced Design," North Atlantic Treaty Organization AGARD Advisory Report 53, prepared by Advisory Group for Aerospace Research and Development, Neuilly Sur Seine, France, 1953, pp. 111–112.

[2] Kaplun, V. A., "Nomographs for Determining the Parameters of Plane Dielectric Layers of Various Structure with Optimum Radio Characteristics," *Radiotechnika i Electronika* (USSR), Part 2, 20(9), 1975, pp. 81–88.

[3] Zamyatin, V. I., A. S. Klyuchnikov, and V. I. Shvets, *Antenna Fairings*, Minsk, USSR: V. I. Lenin Belorussian Publishing House, 1980.

[4] Kaplun, V. A., and V. M. Zelenkevich, "Introduction to Computer-Aided Design of Microwave Antenna Radomes," *Radiotechnika i Electronika (USSR)*, Vol. 42, No. 2, 1987, pp. 89–93.

[5] Mahan, A. I., C. V. Bitterli, and C. G. Wein, "Far Field Diffraction and Boresight Error Properties of a Two Dimensional Wedge," *J. Opt. Soc. Am.*, Vol. 49, No. 6, June 1959, pp. 535–566.

[6] Tricoles, G., "Radiation Patterns and Boresight Error of a Microwave Antenna Enclosed in an Axially Symmetric Dielectric Shell," *J. Opt. Soc. Am.*, Vol. 54, No. 9, September 1964, pp. 1094–1101.

[7] Tricoles, G., "Radiation Patterns of a Microwave Antenna Enclosed in a Hollow Dielectric Wedge," *J. Opt. Soc. Am.*, Vol. 53, No. 5, May 1963, pp. 545–557.

[8] Milligan, T. A., *Modern Antenna Design*, New York: McGraw Hill Book Company, 1985, pp. 215–216.

[9] Kraus, J. D., *Antennas*, 2nd ed., New York: McGraw Hill Book Company, 1988, pp. 174–189.

[10] Kilcoyne, N. R., "An Approximate Calculation of Radome Boresight Error," *Proc. of Technology Symp. on Electromagnetic Windows*, Georgia Institute of Technology, Atlanta, GA, June 1968, pp. 91–111.

[11] Bagby, J., "Desktop Computer Aided Design of Aircraft Radomes," *IEEE MIDCON '88 Conf. Record*, North Hollywood, CA: Western Periodicals Company, 1988, pp. 258–261.

[12] Hayward, R. A., E. L. Rope, and G. Tricoles, "Accuracy of Two Methods for Numerical Analysis of Radome Electromagnetic Effects," *Proc. 14th Symp. on Electromagnetic Windows*, Georgia Institute of Technology, Atlanta, GA, 1978, pp. 53–57.

[13] Hayward, R. A., E. L. Rope, and G. Tricoles, "Accuracy of Two Methods for Numerical Analysis of Radome Electromagnetic Effects," *IEEE Int. Symp. Digest on Antennas and Propagation*, Seattle, WA, June 1979, pp. 598–601.

[14] Raz, S., M. Naor, S. Marcus, and J. Shapira, "Numerical Analysis of Antenna Radome Systems," *Proc. 10th IEEE Convention in Israel*, Tel Aviv, October 1977.

[15] Israel, M., I. Kotlarenko, J. Shapira, and S. Marcus, "A Reference Plane Method for Antenna Radome Analysis," *Proc. 15th Symp. on Electromagnetic Windows*, Georgia Institute of Technology, Atlanta, GA, 1980, pp. 34–39.

[16] Tricoles, G., E. L. Rope, and R. A. Hayward, "Analysis of Radomes by the Method of Moments Method," *Proc. 17th Symp. on Electromagnetic Windows*, Georgia Institute of Technology, Atlanta, GA, 1984, pp. 1–8.

[17] Wu, D. C. F., and R. C. Rudduck, 1974. "Plane Wave Spectrum Surface Integration Technique for Radome Analysis," IEEE *Trans. Antennas Propagation*, AP-22, No. 3, May 1974, pp. 497–500.

[18] Joy, E. B., and G. K. Huddleston, "Radome Effects on Ground Mapping Radar," Final Report, Contract DAAH01-72-C-0598, SF-778 203/0, U. S. Army Missile Command, Huntsville, AL, May 1973.

[19] Huddleston, G. K., H. L. Bassett, and J. M. Newton, "Parametric Investigation of Radome Analysis Methods; Computer Aided Radome Analysis Using Geometric Optics and Lorentz Reciprocity," Final Report, Contract AFOSR-77-3469, Vols. 1–3, Georgia Institute of Technology, Atlanta, GA, 1981.

[20] Joy, E. B., R. E. Wilson, D. E. Ball, and S. D. James, "Comparison of Radome Electrical Analysis Techniques," *Proc. 15th Symp. on Electromagnetic Windows*, Georgia Institute of Technology, Atlanta, GA, 1980, pp. 29–33.

[21] Tavis, M., 1971. "A Three-Dimensional Ray Tracing Method for Calculation of Radome Boresight Error and Antenna Pattern Distortion," TOR-0059(56860), AD 729811, Aerospace Corporation, May 1971.

[22] Einziger, P. D., and L. B. Felson, "Ray Analysis of Two-Dimensional Radomes," *IEEE Trans. Antennas Propagation*, Vol. AP-31, No. 6, 1983, pp. 870–884.

[23] Deschamps, G. A., "Ray Techniques in Electromagnetics," *Proc. IEEE*, Vol. 60, No. 9, September 1972, pp. 1021–1035.

[24] Mahan, A. I., and L. P. Bone, "Far Field Diffraction and Polarization Properties of a Three Dimensional Hollow, Homogeneous, Isotropic Cone," *J. Opt. Soc. Am.*, Vol. 53, No. 5, May 1963, pp. 533–544.

[25] Hayward, R. A., and G. P. Tricoles, "Radome Boresight Error: Numerical Prediction and Physical

Interpretation," *IEEE Antennas and Propagation Society Digest,* IEEE Catalog No. 75CH0963-9AP, June 1975, pp. 61–63.

[26] Einziger, P. D., and L. B. Felsen, "Rigorous Asymptotic Analysis of Transmission Through a Curved Dielectric Slab," *IEEE Trans. Antennas Propagation,* Vol. AP-31, No. 6, November 1983, pp. 863–869.

[27] Richmond, J. H., "Scattering by Dielectric Cylinders of Arbitrary Cross Sectional Shapes," *IEEE Trans. Antennas Propagation,* Vol. AP-13, 1965, pp. 334–341.

[28] Richmond, J. H., "The Calculation of Radome Diffraction Patterns," 1180-13, AD-423660, Department of Electrical Engineering, Ohio State University, September 1966.

[29] Richmond, J. H., "TE Wave Scattering by a Dielectric Cylinder of Arbitrary Cross Sectional Shape," *IEEE Trans. Antennas Propagation,* AP-14, 1966, pp. 460–464.

[30] Tricoles, G., E. L. Rope, and R. A. Hayward, "Wave Propagation Through Axially Symmetric Missile Radomes," Final Report R-81-125, Contract N00019-79-C-0638, AD-A106762/8, 1981.

[31] Paris, D. T., "Computer Aided Radome Analysis," *IEEE Trans. Antennas Propagation,* Vol. AP-18, No. 1, January 1970, pp. 7–15.

[32] Burks, D. G., J. Brand, and E. R. Graf, "The Equivalent Source Concept Applied to the Analysis of Radome Performance," *Proc. Southeastcon '78, Region Three Conf.,* Auburn University, Auburn, AL, April 1978.

[33] Bloom, D. A., P. L. Overfelt, and D. J. White, "Comparison of Spherical Wave Ray Tracing and Exact Boundary Value Solutions for Spherical Radomes," *Proc. 17th Symp. on Electromagnetic Windows,* Georgia Institute of Technology, Atlanta, GA, July 1984.

[34] Kozakoff, D. J., "Geometric Optics Radome Analysis Wall Solution Incorporating the Effects of Wall Curvature," *Proc. SPIE OE/LASE '94 International Symp.,* Los Angeles, CA, January 1994.

[35] Kozakoff, D. J., and J. Hensel, "Materials Implications of Millimeter Wave Radome Performance," *Proc. IEEE Int. Conf. on Infrared (IR) and Millimeter Waves,* Miami, FL, December 1981, pp. 3–5.

[36] White, D. J., and D. J. Banks, "Plane Wave Transmission and Reflection Coefficients for Anisotropic Sheets of Radome Materials," *Proc. 16th Symp. on Electromagnetic Windows,* Georgia Institute of Technology, Atlanta, GA, June 1982, pp. 17–12.

[37] Tricoles, G., and E. L. Rope, "Scattering of Microwaves by Dielectric Slabs and Hollow Dielectric Wedges," *J. Opt. Soc. Am.,* Vol. 55, No. 11, November, pp. 1479–1498.

[38] Ersoy, L., and D. Ford, "RF Performance Degradation Due to Random Radome Surface Irregularities," *IEEE Antennas and Propagation Society Int. Symp. Digest,* 86CH2325-9, 1986, pp. 875–878.

[39] Vorobyev, E. A., "Certain Production Criteria for Large Scale Monolithic Antenna Radomes," *Izvestiya vuz Radioteknika USSR,* Vol. 9, No. 3, 1966, pp. 359–362.

PART III

Computer Implementation

Chapter 7

Radome Modeling Approaches

Parts I and II of this book dealt with theoretical issues; Part III deals with practical issues. How do we apply the theory to solve actual radome-enclosed antenna problems on a PC computer?

This chapter will lay the groundwork in ray-trace modeling issues that are fundamental to all computerized ray-trace techniques discussed previously. In this chapter, you will learn how to apply ray-trace modeling so that you can do the following:

- Mathematically model general radome shapes.
- Understand the vector concepts utilized in ray tracing.
- Find the unit surface normal vector at any arbitrary point on the radome's surface.
- Properly apply radome wall transmission coefficients to an arbitrarily polarized electromagnetic wave.

The organization of the chapter systematically leads you through these issues. Section 7.1 discusses the significance of radome shape and fineness ratio with a particular emphasis on the tangent ogive shape. Section 7.2 presents a method for obtaining the intercept point of an arbitrary ray direction and the radome wall, and Section 7.3 describes techniques for modeling arbitrary (axially symmetric) radome shapes based on polynomial regression.

Section 7.4 discusses methods for computing the angle-of-incidence between a ray and the unit vector surface normal at its intercept point with the radome wall; decomposing an EM wave into components parallel and perpendicular to the plane of incidence defined by the propagation vector and the surface normal, so that the parallel and perpendicular wall transmission coefficients can be applied to the respective wave components; and recombining both wave components to calculate the EM wave on the emergent side of the wall.

In addition, this chapter provides four computer subroutines, designed to help you use the modeling techniques. The software listings are provided for you in Appendixes 7A through 7D at the end of the chapter. The four subroutines are described here:

- Program BRUTE is a straightforward technique to find the coordinates where an arbitrary ray intercepts a radome wall.
- Program ROOTS is a high-speed, algorithm-based rootfinder to locate the coordinates where an arbitrary ray intercepts a radome wall.
- Program POLY is a fifth-order polynomial regression technique that can be used to model an arbitrarily shaped radome.
- Program COMPOSE performs two tasks: (1) decomposes an arbitrarily polarized EM wave at its intercept point into components parallel and perpendicular with respect to the plane of incidence and (2) recombines components on the emergent side of the wall into an emergent wave.

7.1 TANGENT OGIVE RADOME SHAPE

Rationale for Choosing a Particular Shape

For all radome shapes discussed in this section, the outside base diameter D_o and length L_o define the so-called fineness ratio, according to the formula

$$F_o = \frac{L_o}{D_o} \tag{7.1}$$

A similar relationship holds for the inside dimensions L_i and D_i. In this case,

$$F_i = \frac{L_i}{D_i} \tag{7.2}$$

According to Groutage [1], aerodynamic considerations play a major role in radome design, especially in selecting both shape and materials. Shape determines aerodynamic drag, whereas material choice determines the structural response to aerodynamic heating, dynamic pressures, and rain erosion. The decision to reduce drag generally determines the exact shape of the radome. Unfortunately, the most desirable shapes to reduce drag are not the best from the viewpoint of electrical performance.

The potential radome shapes include, but are not limited to, the following: hemisphere, ogive (tangent and secant), Van Karman, and power series. It can be shown that the hemisphere is a special case of the tangent ogive shape, but with a fineness ratio of 0.5. Chin [2] provides an excellent overview of these and other radome shapes, with a complete mathematical description of each.

Weckesser [3] has published a comparison of various radome shapes as shown in Figure 7.1, where the radome is assumed to have a fineness ratio of 2.5. The conical and tangent ogive shapes contain the minimum and maximum volumes, respectively. Figure 7.2 illustrates the aerodynamic drag, showing the approximate drag coefficient for the four shapes. The tangent ogive shape has the worst drag, whereas the secant ogive shape has the best.

However, Yost et al. [4] point out that the fineness ratio has a much greater and pronounced effect on drag coefficient than shape. For instance, these authors cite a particular missile for which a 28% increase in fineness ratio decreased the drag coefficient by 35% at Mach 3. The reduction in drag coefficient due to a potential change in shape was very small.

An analytical study by Crowe [5] indicated that, for fineness ratios of less than 2.25, the secant ogive radome causes less BSE than either the Von Karman, tangent ogive, or the power series radome of the same fineness ratio. However, for larger fineness ratios, the tangent ogive shape is best.

For the purpose of radome modeling, in this book we have emphasized the tangent ogive radome shape for the following reasons:

(a)

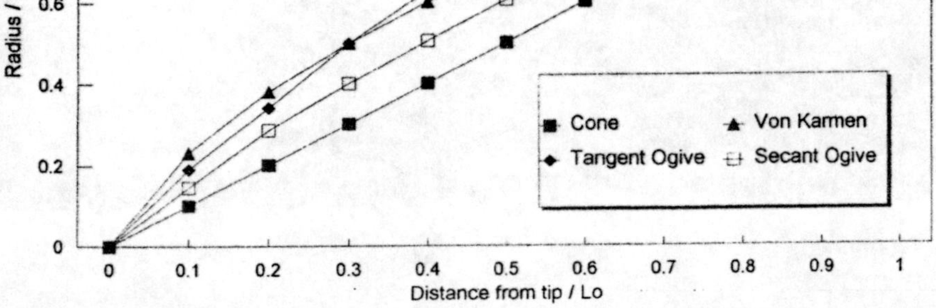

Figure 7.1 (a) A comparison of some common radome shapes and (b) experimental tactical missile radomes with different nose configurations. (Photo courtesy of U.S.A.F. Wright Laboratory Armament Division.)

Figure 7.1 (continued)

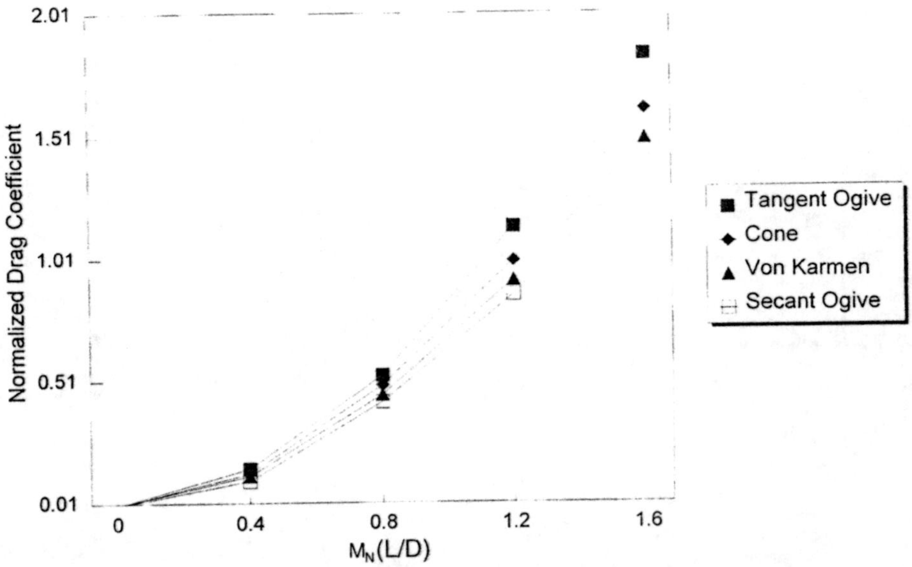

Figure 7.2 Wave drag of common radome shapes. (M_N is the mach number.)

- Relatively easy mathematical modeling;
- Representative of many common radomes, ranging from a hemisphere (fineness ratio = 0.5) to very streamlined radomes (fineness ratios > 3);
- Mathematical principles that can be transferred to other detailed radome shapes.

The performance of a radome, in general, degrades very rapidly as the maximum angle of incidence exceeds about 75 degrees. For a tangent ogive shape, Figure 7.3 illustrates the relationship between the maximum incidence angle for an on-axis target and the fineness ratio of the radome. (An on-axis target produces worst-case incidence angles.) From this plot, it is clear that the maximum incidence angle can only be reduced by a decrease in fineness ratio. From an electrical viewpoint, a hemisphere is most desirable because of very small incidence angles and resulting small BSE effects.

Mathematical Modeling of Shape

Figure 7.4 defines the key parameters needed to describe mathematically a tangent ogive radome shape. Any point on the outside surface must satisfy the following equation:

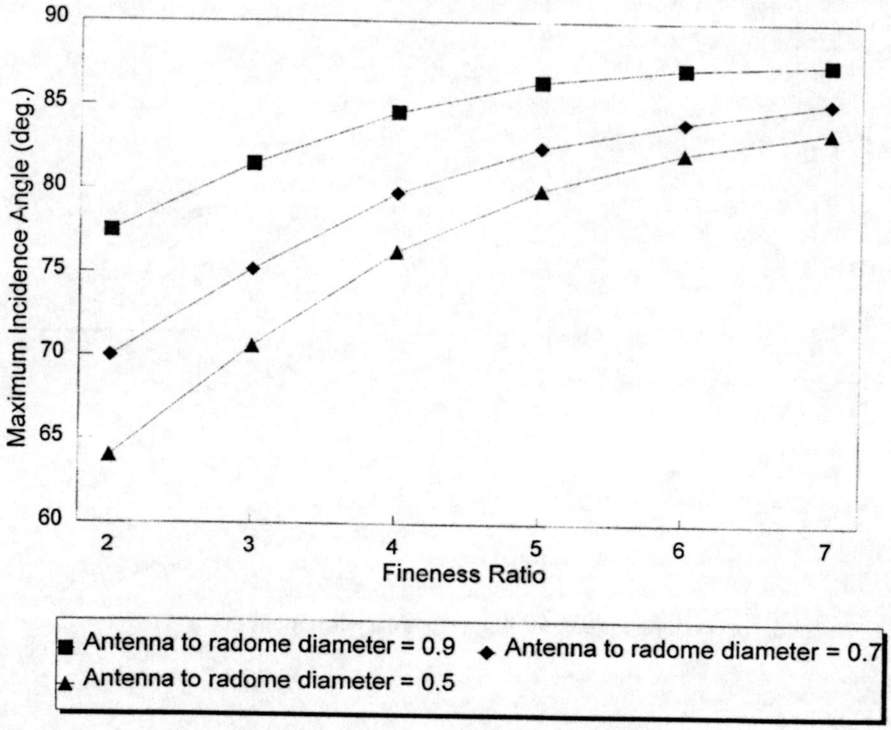

Figure 7.3 Maximum incidence angles for on-axis antenna pointing (tangent ogive radome).

$$(r_p + B)^2 + x_p^2 = R^2 \tag{7.3}$$

where this equation applies for fineness ratios of 0.5 or larger.

If the x-axis corresponds to the radome axis, then the radome shape is a surface of revolution formed by revolving a tangent ogive curve about the x-axis. The cross sections through this three-dimensional shape are circular in the y-z plane with the radome radius given by

$$r_p = \sqrt{z_p^2 + y_p^2} \tag{7.4}$$

From Figure 7.4, we clearly see that

$$R^2 = L_o^2 + B^2 \tag{7.5}$$

$$\frac{D_o}{2} = R - B \tag{7.6}$$

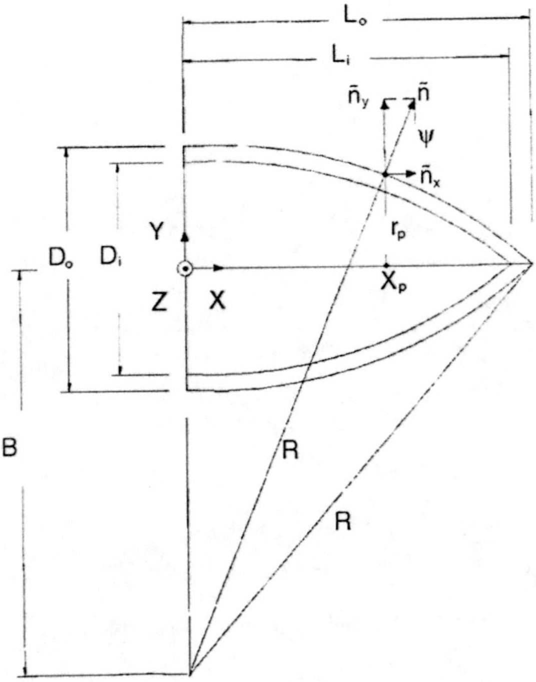

Figure 7.4 Tangent ogive radome geometry.

By simultaneous solution of these two equations, we obtain expressions for the two unknowns in terms of the radome length and diameter:

$$B = \frac{4L_o^2 - D_o^2}{4D_o} \tag{7.7}$$

$$R = \frac{4L_o^2 + D_o^2}{4D_o} \tag{7.8}$$

By substitution of (7.7) and (7.8) into (7.3), we can obtain the formula of the tangent ogive shape in terms of known constants L_o and D_o.

Computation of Surface Normal Vector

For any point $P(x_p, y_p, z_p)$ on the tangent ogive contour, where $y = r_p$, the slope of the curve is found from

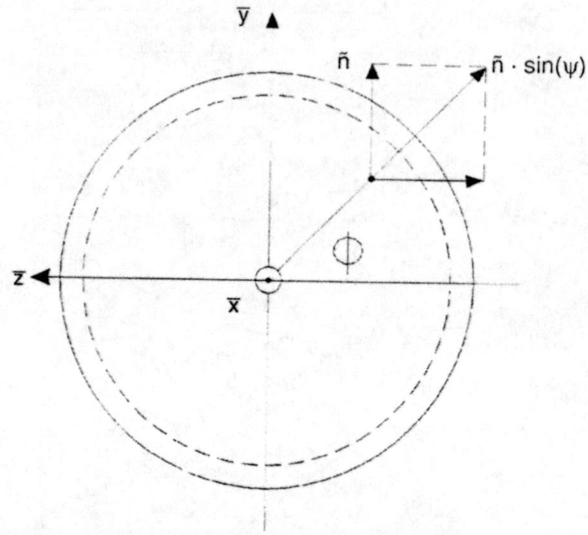

Figure 7.5 Components of the surface normal vector in the *x-y* plane.

$$m_s = \frac{dy}{dz} = \frac{-x_p}{r_p + B} \tag{7.9}$$

Therefore, the slope of the surface normal vector, $\bar{n}$, also illustrated in Figure 7.4, is given by the formula

$$\tan(\Psi) = \frac{-1}{m_s} = \frac{r_p + B}{x_p} \tag{7.10}$$

The *x*-component of the surface normal vector is

$$n_x = \cos(\Psi) = \frac{x_p}{\sqrt{(r_p + B)^2 + x_p^2}} \tag{7.11}$$

At a ray intercept point $P(x_p, y_p, z_p)$, illustrated in Figure 7.5, the *y* and *z* components of the surface normal vector are found from

$$n_z = \sin(\Psi)\cos(\Phi) = \frac{r_p + B}{\sqrt{(r_p + B)^2 + x_p^2}} \frac{z_p}{r_p} \tag{7.12}$$

and

$$n_y = \sin(\Psi)\sin(\Phi) = \frac{r_p + B}{\sqrt{(r_p + B)^2 + x_p^2}} \frac{y_p}{r_p} \tag{7.13}$$

7.2 DETERMINING THE INTERCEPT POINT OF A RAY WITH A TANGENT OGIVE RADOME WALL

All ray-trace radome analysis methods require the determination of the point-of-intersection of a ray and the radome surface [6]. Some radome shapes, such as the cone, lend themselves to an exact solution of the intercept point. Other shapes, such as the Von Karman, do not [7]. This subsection describes three methods of determining the intercept point.

Method 1

Huddleston and Balius [8] introduced a method for finding the intercept point that can apply to the tangent ogive radome geometry. Using this method, the outer radome shape can be represented by samples spaced equally along the radome axis. Any two sample points are sufficient to define a cone with the vertex beyond the tip and opening toward the radome base. Between the sample points, the intersection of an incident ray and the cone approximates the point of intersection with the actual surface. The intersection point is then refined, using additional sample points and a perturbational technique.

Method 2

Joy and Ball [9] also reported development of a ray intercept technique. They specified the radome surface as cylindrical radius values for constant station (z position) values. This method of radome shape specification is typical of radome shape data available to the radome analyst.

The technique is an interactive algorithm that locally approximates the radome surface as a cone that is tangent to the radome surface at the point of the ray intercept point. Using the interactive algorithm, the analyst can iteratively vary the fineness ratio and the location of the cone axis so that the intersection point of the ray and the cone equals the intersection point of the ray and the radome surface. Convergence is always possible for monotonically concave radome shapes.

Method 3

This method of determining the intercept point is, like Method 2, an iterative method. However, it is more general than the two previously discussed methods and is particularly well suited to computer use. This technique uses a generalized ray-tracing method for tangent ogive shapes that are treated as surfaces of revolution formed by revolving a single-valued plane curve about the radome axis.

Referring to Figure 7.6(a), the problem starts with definition of a vector originating at a point $P(x_o, y_o, z_o)$ within the radome and having a propagation direction $\underline{k}$. The coordinates corresponding to the tip of the vector can be expressed as a parametric equation via [10]:

$$x_i = x_0 + C_i k_x \tag{7.14}$$

$$y_i = y_0 + C_i k_y \tag{7.15}$$

$$z_i = z_0 + C_i k_z \tag{7.16}$$

where k_x, k_y, k_z, are the direction cosines of the line, and C_i is the distance along the line. There is a unique value of C_i at the intercept point of interest. The point $P(x_o, y_o, z_o)$ is arbitrary in this example, but in actual radome analysis, it will correspond to a point on the antenna aperture surface.

As we iteratively vary the vector length, we should think of the vector as a ray. The procedure for finding the intersection of this ray with the radome surface consists of varying the parameter C_i iteratively until we learn the intercept point. At the intercept point, the coordinates of the tip of the vector will correspond to a point on the radome surface, as shown in Figure 7.6(b). Usually, C_i will fall in the range of $0 < C_i < L_o$, where L_o is the radome length.

The process is aided by definition of an error function ERF, which is the difference between r_p and r_i:

$$\text{ERF} = r_p - r_i = \left(\sqrt{R^2 - x_i^2} - B\right) - \sqrt{z_i^2 + y_i^2} \tag{7.17}$$

Clearly, ERF will be positive for a value of C_i that is too small, and negative for a value of C_i that is too large. It will be exactly zero only for the correct intercept point. Figures 7.6(a–c) illustrate the conditions where C_i is too small, too large, or the correct value, respectively.

When carrying out this procedure in a computerized radome analysis code, an algorithm that finds the zero crossing of ERF speeds up the process of determining the correct value of C_i. Figure 7.7 shows one of the simplest rootfinder methods in the software flow diagram. The approach increments coarse, then medium, and finally fine increments in C_i values, until detecting the zero crossing in ERF value. These increments correspond to 1.0, 0.1, and 0.01 in., respectively. Once we learn the correct intercept point, we can evaluate the surface normal vector at the intercept point via (7.11) through (7.13).

Appendix 7A is a sample computer program listing that shows this particular rootfinder. This program incorporates a subroutine that models the tangent ogive radome shape discussed in this chapter. The development works for tangent ogive radomes having hemispherical shape, $L_o/D_o = 0.5$, or higher fineness ratios. Typically, this program takes from 10 to 25 cycles to find the intercept point to within an accuracy of 0.0254 cm.

The use of a high-speed rootfinder algorithm can reduce the computer run

(a)

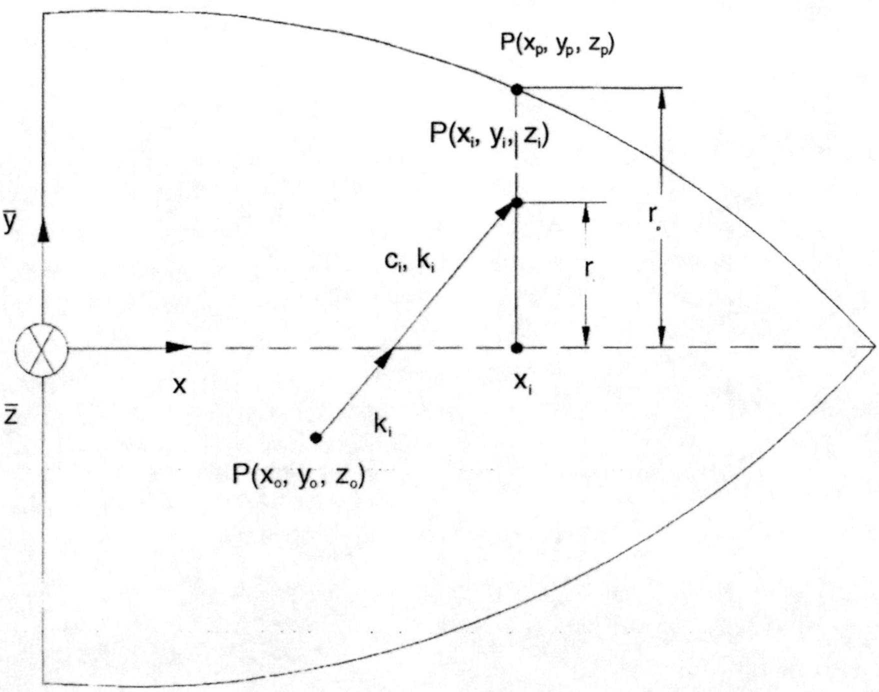

Figure 7.6 Three cases of vector intercept with radome wall: (a) vector tip inside radome, (b) vector tip intercepting radome surface, and (c) vector tip outside radome.

time to detect the root of ERF, and therefore the correct value of C_i. Figure 7.8 shows the computer software flow diagram for this approach. Here, we treat C_i and ERF as parameters with an index *I*. Figure 7.9 illustrates the iterative steps of this algorithm. Referring to Figure 7.9(a), the approach is as follows:

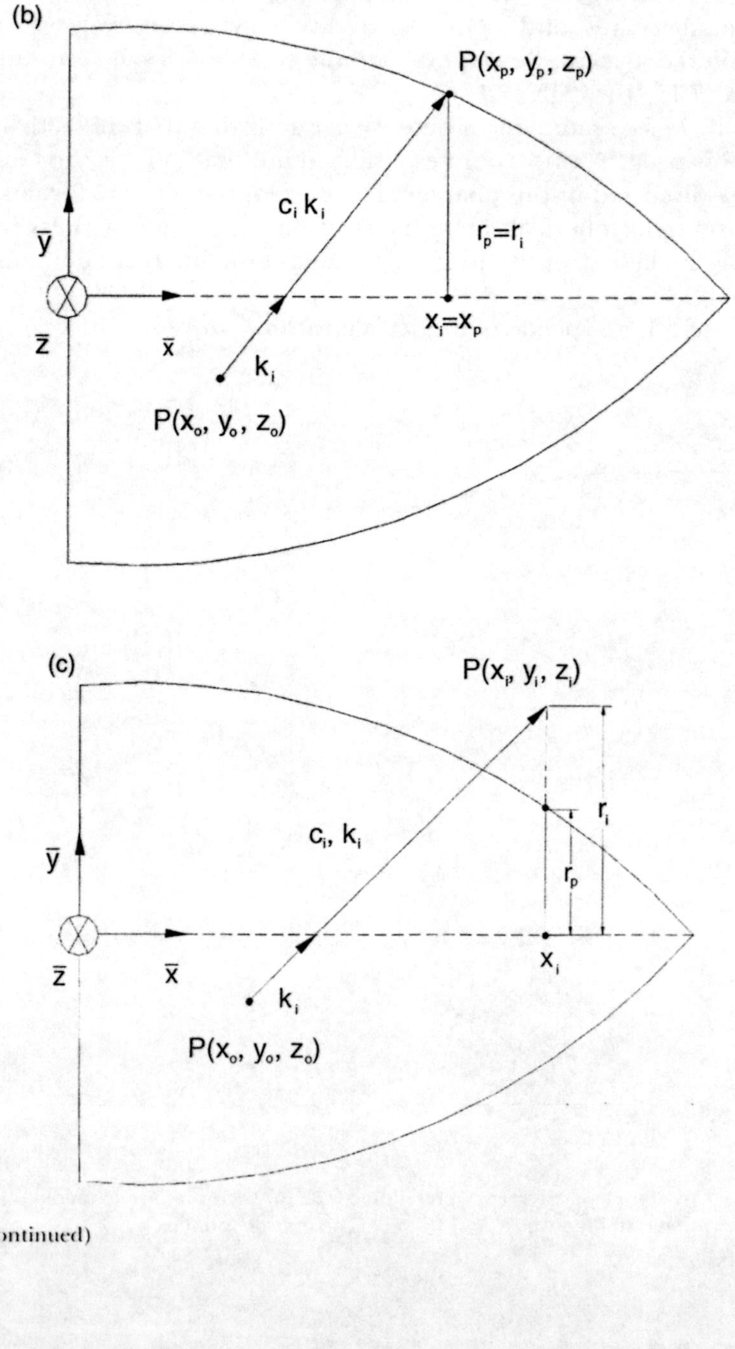

Figure 7.6 (continued)

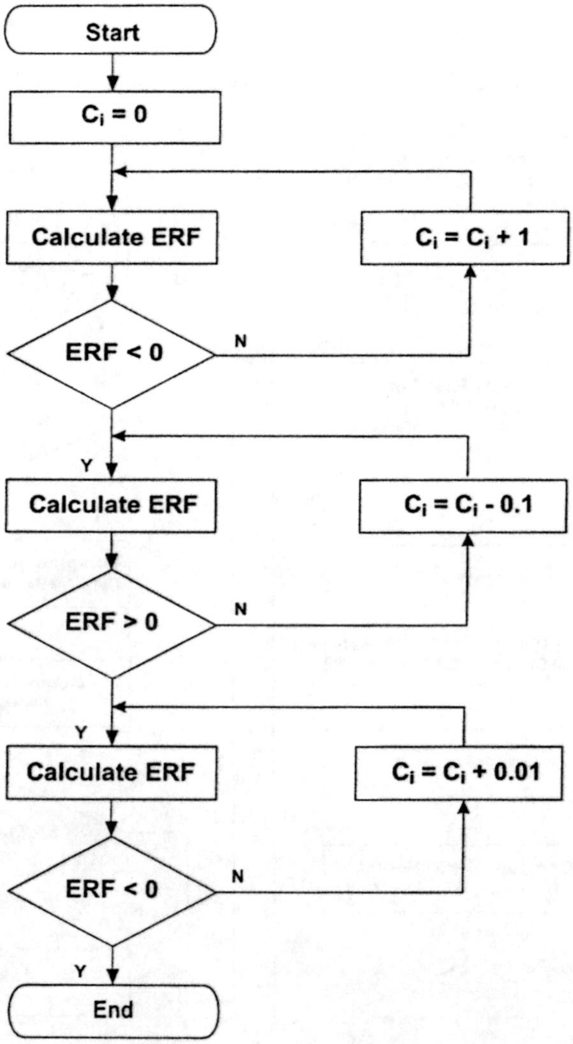

Figure 7.7 Computer software flow diagram for Program BRUTE rootfinder algorithm.

1. Assume that the function is a straight line with a slope m. Its values for the extremes, $C_i = 0$ and $C_i = L_o$, are determined; these are F and G, respectively.
2. Determine the initial root of this linear relationship, $C_i(0)$, and the function value ERF(0).
3. Define a line with the same slope m, but passing through ERF(0).
4. Compute its zero crossing $C_i(1)$, and then the accompanying value of ERF(1).

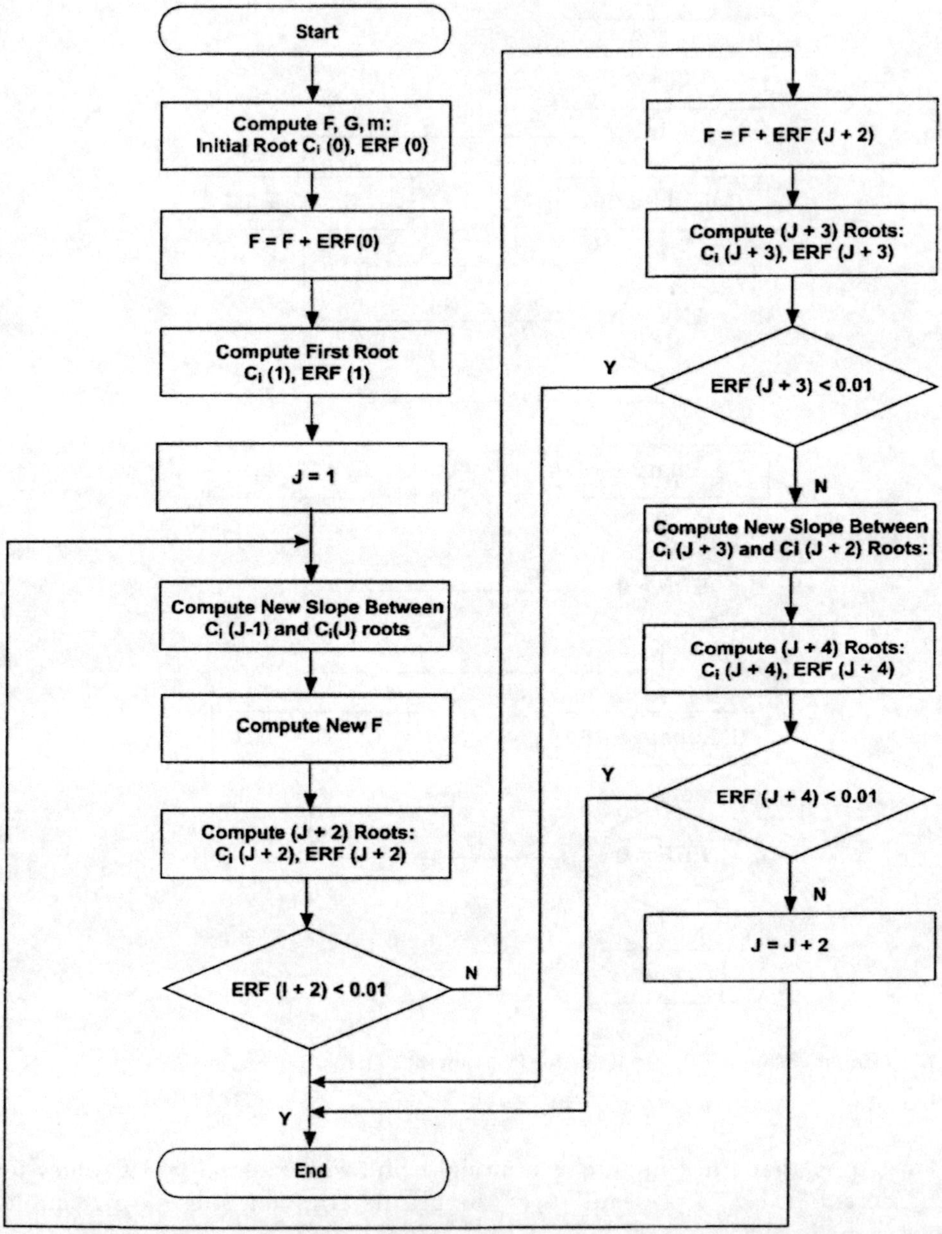

Figure 7.8 Computer software flow diagram for Program ROOTS high-speed rootfinder algorithm.

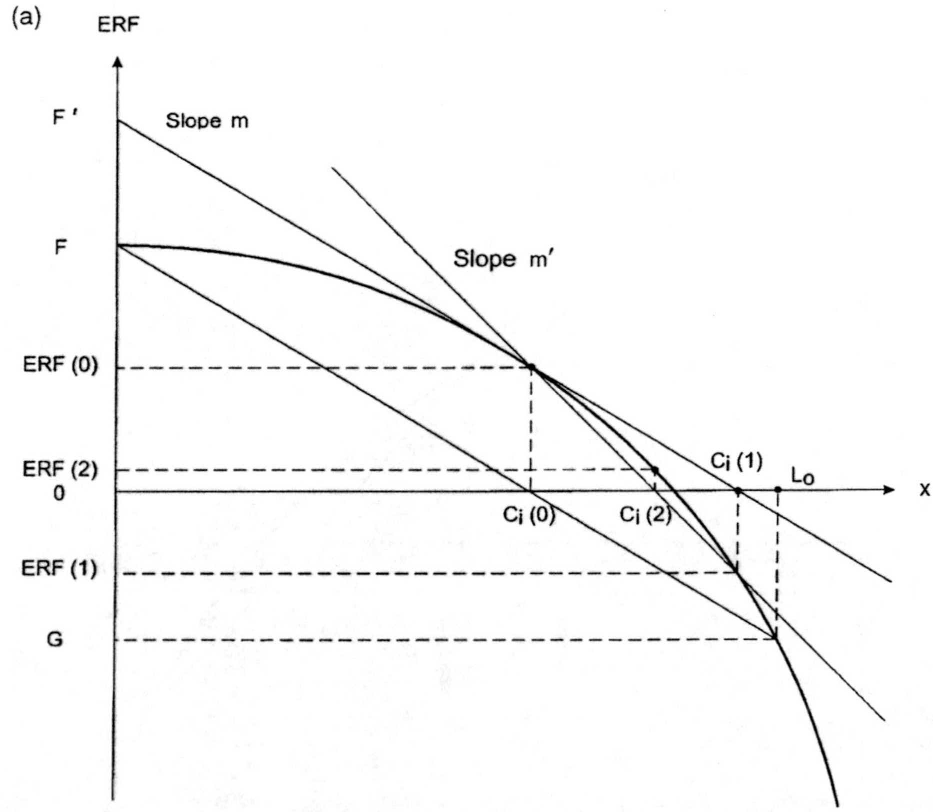

Figure 7.9 Illustration of iterative steps in high-speed rootfinder algorithm: (a) determination of initial roots and (b) determination of higher roots.

5. Define a new slope, m', passing through the points $\{C_i(0), \text{ERF}(0)\}$ and $\{C_i(1), \text{ERF}(1)\}$. The zero crossing of this new line is $C_i(2)$, and the value of the function is ERF(2). If $\text{ERF}(2) < 0.01$, select $C_i(2)$ as the root and compute the coordinates of the intercept point via (7.14) through (7.16). If $C_i(2)$ is not the root, take it as the initial root and repeat all earlier steps, as illustrated in Figure 7.9(b). These computations result in determination of roots $C_i(2)$, $C_i(3)$, and $C_i(4)$.

Appendix 7B is a computer program software listing for a high-speed root-finding algorithm based on the technique described earlier. The development works for tangent ogive radomes having a fineness ratio of 0.5 or greater. This method is 5 to 10 times faster than the brute force technique employed in Appendix 7A. You can use the software to develop new computerized radome analysis codes.

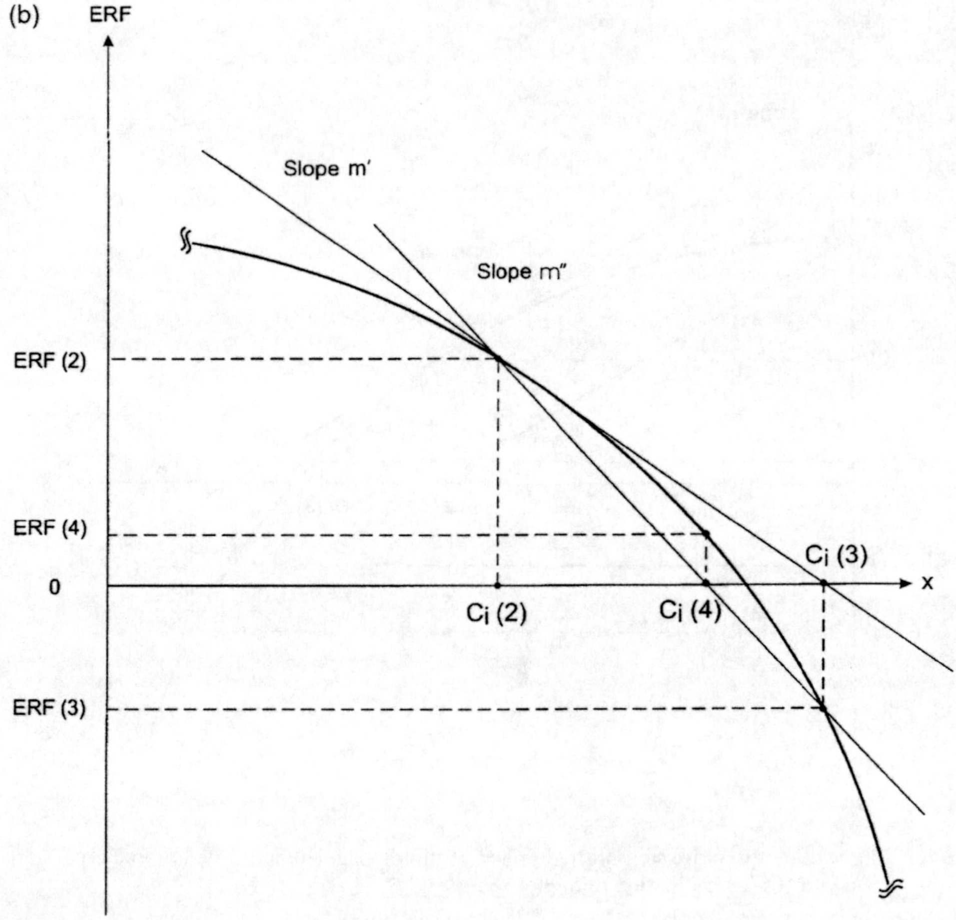

Figure 7.9 (continued)

7.3 INTERCEPT OF RAY WITH GENERAL RADOME SHAPES HAVING AXIAL SYMMETRY

This section discusses radomes whose shapes are not easily described mathematically in a closed-form solution, such as the tangent ogive radome shape. However, if the radius (r_p) is known as a function of distance from the radome base (x_p), then the data can be fitted to a polynomial equation using the least squares method, with the following form:

$$r_p = A_0 + A_1 x_p + A_2 x_p^2 + A_3 x_p^3 + A_4 x_p^4 + A_5 x_p^5 \tag{7.18}$$

To find a ray intercept point and surface normal vector for a tangent ogive radome, this form can be used with the previously developed Method 3 of Section 7.2. To apply this method, we must use radome shape data to evaluate the unknown constants in (7.18).

Solving the following system of equations can calculate the regression coefficients A_i. Gauss's elimination method with partial pivoting can be used.

$$\begin{bmatrix} N & \sum x_p & \sum x_p^2 & \sum x_p^3 & \sum x_p^4 & \sum x_p^5 \\ \sum x_p & \sum x_p^2 & \sum x_p^3 & \sum x_p^4 & \sum x_p^5 & \sum x_p^6 \\ \sum x_p^2 & \sum x_p^3 & \sum x_p^4 & \sum x_p^5 & \sum x_p^6 & \sum x_p^7 \\ \sum x_p^3 & \sum x_p^4 & \sum x_p^5 & \sum x_p^6 & \sum x_p^7 & \sum x_p^8 \\ \sum x_p^4 & \sum x_p^5 & \sum x_p^6 & \sum x_p^7 & \sum x_p^8 & \sum x_p^9 \\ \sum x_p^5 & \sum x_p^6 & \sum x_p^7 & \sum x_p^8 & \sum x_p^9 & \sum x_p^{10} \end{bmatrix} \begin{bmatrix} A_0 \\ A_1 \\ A_2 \\ A_3 \\ A_4 \\ A_5 \\ A_5 \end{bmatrix} = \begin{bmatrix} \sum r_p \\ \sum x_p r_p \\ \sum x_p^2 r_p \\ \sum x_p^3 r_p \\ \sum x_p^4 r_p \\ \sum x_p^5 r_p \end{bmatrix} \tag{7.19}$$

Appendix 7C lists a software code that solves this system of equations and computes the unknown coefficients. The next section describes the procedure for using this program.

Basic Information About the POLY Program

This program listing is in BASIC. Before running the program, create a table with the $\{x_p, r_p\}$ data sets that describe the radome. The program will do the least squares fit regression analysis to ensure the approximation is as close as possible over the entire range of the data. The number of data points that should be entered to get a good approximation is something of a trial-and-error process. Too few points may be insufficient for an accurate approximation, whereas too many are tedious to enter and will not significantly improve the data.

To run the program, perform the following steps:

1. First enter the x_p values and then the r_p values for the first data pair, as you are prompted. When entering values, you may edit them with the backspace key before you press [Return]. You can abort execution at any point by pressing [Ctrl-Break].
2. Enter the rest of the data sets as the program prompts you.
3. Type E to "exit" the data entry screen after you have entered the final pair.
4. Wait for the program to run the regression analysis, ending by the display of the coefficients, A_0 through A_5, which provide the best least squares approximation to your data.

5. Enter a value for x_p when prompted to test this approximation by allowing you to see what value of r_p results.
6. Enter Q for "quit." The program will ask if you want to try the regression analysis again with a different set of input data. Enter Y (yes) or N (no) as required.

Determining Intercept Point and Surface Normal

Once the correct coefficients are determined, you can locate a ray intercept point by iterating C_i to find where the error function ERF is zero by using the rootfinder techniques of Section 7.2. However, the error function must be expressed in terms of the polynomial coefficients:

$$\text{ERF} = r_p - r_i = \left(A_0 + A_1 x_p + A_2 x_p^2 + A_3 x_p^3 + A_4 x_p^4 + A_5 x_p^5\right) - \sqrt{z_p^2 + y_p^2} \tag{7.20}$$

In this case, we must iterate the values of x_p to find the zero of the error function. Once the correct value of x_p is determined, the slope passing through the point x_p is found from

$$m_s = \frac{dy}{dz} = A_1 + 2A_2 x_p + 3A_3 x_p^2 + 4A_4 x_p^3 \tag{7.21}$$

The slope of the surface normal vector is

$$\tan(\Psi) = \frac{-1}{m_s} = \frac{-1}{A_1 + 2A_2 x_p + 3A_3 x_p^2 + 4A_4 x_p^3} \tag{7.22}$$

We can resolve the three components of the normal vector following (7.11) through (7.13).

7.4 Wave Decomposition at Intercept Point

The flat-plate voltage transmission coefficients of the radome wall must weight the incident electric field associated with each ray. The problem geometry is illustrated in Figure 7.10, which shows an arbitrarily polarized wave E^i incident on a dielectric wall at angle θ with respect to the unit surface normal at the intercept point, $\bar{n}$. The assumptions are that (1) the wave is a plane wave consistent with ray tracing and (2) the wall will be treated as locally plane at the ray intercept point.

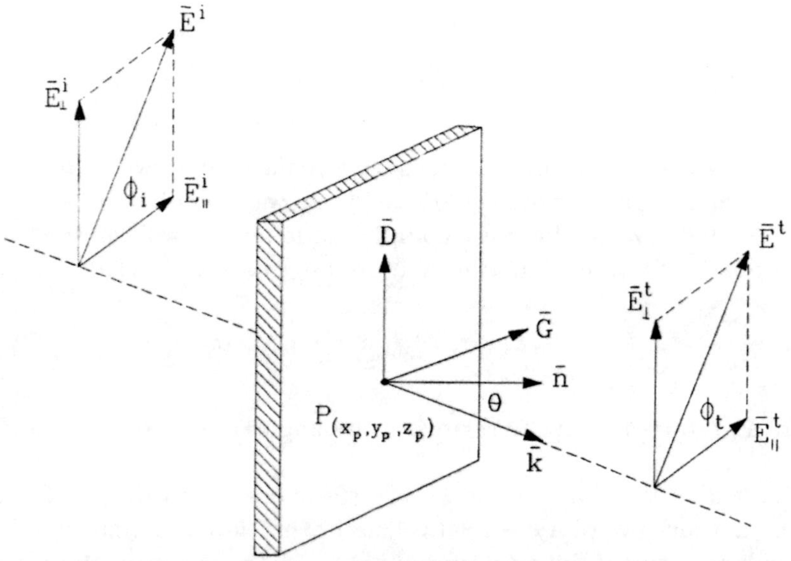

Figure 7.10 Decomposition of an incident wave into parallel and perpendicular components.

To apply transmission coefficients to the wave as it propagates through the wall, we must first resolve the arbitrarily polarized incident wave into components parallel and perpendicular to the plane of incidence.

Decomposition of the incident electric field into two orthogonal vector components follows from vector mathematics. First, it is necessary to determine unit vectors perpendicular and parallel to the plane formed by the surface normal vector, $\bar{n}$, and the direction of propagation vector, $\bar{k}$. The perpendicular vector is given by [11,12]

$$\bar{D} = \bar{n} x \bar{k} = \frac{(n_y k_z - n_z k_y)\bar{x} + (n_z k_x - n_x k_z)\bar{y} + (n_x k_y - n_y k_x)\bar{z}}{\sqrt{(n_y k_z - n_z k_y)^2 + (n_z k_x - n_x k_z)^2 + (n_x k_y - n_y k_x)^2}} \tag{7.23}$$

or

$$\bar{D} = d_x \bar{x} + d_y \bar{y} + d_z \bar{z} \tag{7.24}$$

and the parallel vector by

$$\bar{G} = \bar{D} x \bar{k} = \frac{(d_y k_z - d_z k_y)\bar{x} + (d_z k_x - d_x k_z)\bar{y} + (d_x k_y - d_y k_z)\bar{z}}{\sqrt{(d_y k_z - d_z k_y)^2 + (d_z k_x - d_x k_z)^2 + (d_x k_y - d_y k_z)^2}} \tag{7.25}$$

or

$$\overline{G} = g_x\overline{x} + g_y\overline{y} + g_z\overline{z} \tag{7.26}$$

We can also represent any transverse electromagnetic wave as two orthogonally polarized waves. Let E^i represent the electric field of a wave incident on the wall, as in Figure 7.10. Since this vector field is a complex function, we will represent the incident field in phasor notation similar to that used by Hollis et al. [13]:

$$\overline{E}^i\angle\sigma_i = E_x^i\angle\sigma_x\overline{x} + E_y^i\angle\sigma_y\overline{y} + E_z^i\angle\sigma_z\overline{z} \tag{7.27}$$

where σ_x, σ_y, and σ_z represent the relative phase angle of each of the three components of E^i.

For this example, assume that we are sufficiently into the far field of the source antenna where the phase is relatively uniform. Note also that right-hand circular polarization occurs if the y component has a net −90-degree phase shift compared with the horizontal component. It is left-hand circular if the y component has a net −90-degree phase shift compared with the horizontal component. The Institute of Electrical and Electronic Engineers' standards define the sense to be right hand if the electric vector rotates in a clockwise direction for an observer looking in the direction of propagation. The sense is right-hand circular if the electric vector rotates in the opposite direction.

As illustrated in Figure 7.10, the incident vector field is resolved into scalar components_perpendicular and parallel to the plane of incidence by vector dot products of E_i, respectively. The result obtained is:

$$E_\perp^i\angle\sigma_\perp^i = E_x^i d_x\angle\sigma_x + E_y^i d_y\angle\sigma_y + E_z^i d_z\angle\sigma_z \tag{7.28}$$

$$E_\parallel^i\angle\sigma_\parallel^i = E_x^i g_x\angle\sigma_x + E_y^i g_y\angle\sigma_y + E_z^i g_z\angle\sigma_z \tag{7.29}$$

The transmitted field is expressed as the product of the incident field and the complex voltage transmission coefficients for each of the two orthogonal linear polarizations. Therefore, the magnitude and phase of each of the two components of the transmitted field are:

$$E_\perp^t\angle\sigma_\perp^t = E_\perp^i T_{W\perp}\angle(\sigma_\perp^i + \mathrm{IPD}_\perp) \tag{7.30}$$

$$E_\parallel^t\angle\sigma_\parallel^t = E_\parallel^i T_{W\parallel}\angle(\sigma_\parallel^i + \mathrm{IPD}_\parallel) \tag{7.31}$$

where $T_{W\perp}$ and $T_{W\parallel}$ are the voltage transmission coefficient for the perpendicular and parallel wave components, respectively (dimensionless), and $\mathrm{IPD}_\perp$ and $\mathrm{IPD}_\parallel$ are the insertion phase delays for the perpendicular and parallel wave components, respectively (radians).

We can now recombine vector components to obtain the transmitted vector field:

$$\overline{E}_t \angle \sigma_t = (E^t_\perp d_x \angle \sigma^t_\perp + E^t_\parallel g_x \angle \sigma^t_\parallel)\overline{x} + (E^t_\perp d_y \angle \sigma^t_\perp + E^t_\parallel g_y \angle \sigma^t_\parallel)\overline{y} + (E^t_\perp d_z \angle \sigma^t_\perp + E^t_\parallel g_z \angle \sigma^t_\parallel)\overline{z} \quad (7.32)$$

To demonstrate this technique, Appendix 7D shows a sample computer program for an electromagnetic wave decomposition and recomposition. The program breaks the wave into parallel and perpendicular components, applies the complex, plane wave, wall transmission coefficients, and returns a complex result for the transmitted electric field.

References

[1] Groutage, F. D., "Radome Development for a Broadband RF Missile Sensor," Report NELC-TR-2023, Naval Electronics Laboratory Center, San Diego, CA, January 1977.

[2] Chin, S. S., *Missile Configuration Design*, New York: McGraw Hill Book Company, 1981.

[3] Weckesser, L. B., "Thermal-Mechanical Design Principles," Chap. 3 in *Radome Engineering Handbook*, Walton, J. D., Jr., Ed., New York: Marcel Dekker, 1970.

[4] Yost, D. J., L. B. Weckess, and R. C. Mallalieu, "Technology Survey of Radomes for Anti-Air Homing Missiles," Report FS 80-022, Applied Physics Laboratory, John Hopkins University, Laurel, MD, March 1980.

[5] Crowe, B., *Air Launched Tactical Missile Radome Study*, Newport Beach, CA: Flight Systems, January 1977.

[6] Joy, E. B., and G. K. Huddleston, *Radome Effects on the Performance of Ground Mapping Radar*, Final Report, Contract DAAH01-72-C-0598, Georgia Institute of Technology, Atlanta, GA, March 1973.

[7] Walton, J. D., Ed., *Radome Engineering Handbook*, New York: Marcel Decker, 1970.

[8] Huddleston, G. K., and A. R. Balius, "A Generalized Ray Tracing Method for Single Valued Radome Surfaces of Revolution," *Proc. 15th Symp. on Electromagnetic Windows*, Georgia Institute of Technology, Atlanta, GA, June 1980.

[9] Joy, E. B., and D. E. Ball, "A Fast Ray Tracing Algorithm for Arbitrary Monotonically Concave Three-Dimensional Radome Shapes," *Proc. 17th Symp. on Electromagnetic Windows*, Georgia Institute of Technology, Atlanta, GA, July 1984.

[10] Adams, L. J., and P. A. White, *Analytic Geometry and Calculus*, London: Oxford University Press, 1968, p. 542.

[11] Kreyszig, E., *Advanced Engineering Mathematics*, New York: John Wiley and Sons, 1962.

[12] Pipes, L. A., and L. R. Harvell, *Applied Mathematics for Engineers and Physicists*, 3rd ed., New York: McGraw Hill Book Company, 1970.

[13] Hollis, J. S., T. J. Lyon, and L. Clayton, Jr., *Microwave Antenna Measurements*, 2nd ed., Atlanta, GA: Scientific Atlanta, 1970.

Appendix 7A: Program BRUTE Software Listing

CLS

Sample program to demonstrate brute force rootfinder.

Works for L0/D0 = 0.5, or higher, fineness ratio tangent ogive radome shapes. (In this sample program, radome axis is the $\bar{z}$ axis.)

Variable List

L0	= radome length (inches)
D0	= radome base diameter (inches)
R,B	= parameters used to describe tangent ogive geometry (inches)
x0,y0,z0	= initial point within radome, in standard cartesian c oordinates (inches)
xi,yi,zi	= a general point along the line between the initial point and the intercept point (inches)
xp,yp,zp	= intercept point on the radome outer surface, in standard cartesian coordinates (inches)
kx,ky,kz	= unit propagation vector (inches ^ -1)
Ci	= convergence constant (dimensionless)
nx,ny,nz	= unit surface normal vector at intercept point (inches)
ERF	= computational error function (inches)

```
Enter Data Here:
  COLOR 15, 1
  CLS

     INPUT "ENTER RADOME LENGTH (inches) = "; L0
     INPUT "ENTER RADOME BASE DIAMETER (inches) = "; D0
     PRINT "":  PRINT ""
     PRINT "ENTER INITIAL POINT WITHIN RADOME"
     PRINT ""
        INPUT "x0 ="; x0
        INPUT "y0 ="; y0
        INPUT "z0 ="; z0
     PRINT "":     PRINT ""
   PRINT "INPUT PROPAGATION VECTOR"
        INPUT "kx ="; kx
        INPUT "ky ="; ky
        INPUT "kz ="; kz

  '  Begin Calculations Here:

' One Time Calculation for Radome Parameters R,B:
        B = (4 * L0 ^ 2 - D0 ^ 2) / (4 * D0)
        R = (4 * L0 ^ 2 + D0 ^ 2) / (4 * D0)

' BRUTE FORCE ROOTFINDER TO FIND Ci STARTS HERE:
      Ci = 0
'       First: Increment Ci in inches:
```

```
        FOR I = 1 TO 1000
          Ci = Ci + 1
          xi = x0 + Ci * kx
          yi = y0 + Ci * ky
          zi = z0 + Ci * kz
          GOSUB 100
          IF ERF = 0 GOTO 200
          IF ERF < 0 THEN EXIT FOR
        NEXT I

'       Second: Increment Ci in tenths of inches:
        FOR I = 1 TO 10
          Ci = Ci - .1
          xi = x0 + Ci * kx
          yi = y0 + Ci * ky
          zi = z0 + Ci * kz
          GOSUB 100
          IF ERF = 0 THEN GOTO 200
          IF ERF > 0 THEN EXIT FOR
        NEXT I

'       Third: Increment Ci in hundredths of inches:
        FOR I = 1 TO 100
          Ci = Ci + .01
          xi = x0 + Ci * kx
          yi = y0 + Ci * ky
          zi = z0 + Ci * kz
          GOSUB 100
          IF ERF = 0 THEN GOTO 200
          IF ERF < .01 THEN EXIT FOR
      NEXT I

' Calculate Normal Vector at Ray Intercept Point:

200  '  Continue

        xp = xi: yp = yi: zp = zi
        rp = SQR(xp ^ 2 + yp ^ 2)
        nx = ((rp + B) / SQR((rp + B) ^ 2 + zp ^ 2)) * xp / rp
        ny = ((rp + B) / SQR((rp + B) ^ 2 + zp ^ 2)) * yp / rp
        nz = zp / SQR((rp + B) ^ 2 + zp ^ 2)
    PRINT "":   PRINT ""

  PRINT "RAY INTERCEPT POINT IS: xp ="; xp; " yp = "; yp; " zp = "; zp
  PRINT "SURFACE NORMAL IS: nx = "; nx; " ny = "; ny; " nz = "; nz

END

100  '  SUBROUTINE TANGENT OGIVE SHAPE

        'to prevent overflows:
        IF xi = 0 AND yi = 0 THEN xi = 1E-12
```

```
        IF zi <=  L0 THEN
          ERF = (SQR(R ^ 2 - zi ^ 2) - B) - SQR(xi ^ 2 + yi ^ 2)
        END IF
        IF z > L0 THEN
          ERF = -(SQR(R ^ 2 - (2 * L0 - zi) ^ 2) - B) -SQR(xi ^ 2
+ yi ^ 2)
        END IF

RETURN
```

Appendix 7B: Program ROOTS Software Listing

Code to demonstrate high speed root finder

Works for L0/D0 = 0.5, or higher, fineness ration tangent ogive radome shapes.

CLS

DIM ERF(30), Ci(30)

(In this sample program, radome axis is the $\bar{z}$ axis.)

Variable List

L0	= radome length (inches)
D0	= radome base diameter (inches)
R,B	= parameters used to describe tangent ogive geometry (inches)
x0,y0,z0	= initial point within radome, in standard cartesian c oordinates (inches)
xi,yi,zi	= a general point along the line between the initial point and the intercept point (inches)
xp,yp,zp	= intercept point on the radome outer surface, in standard cartesian coordinates (inches)
kx,ky,kz	= unit propagation vector (inches ^ -1)
Ci	= convergence constant (dimensionless)
nx,ny,nz	= unit surface normal vector at intercept point (inches)
ERF	= computational error function (inches)

Enter Data Here:

```
   COLOR 15, 1
   CLS

     INPUT "ENTER RADOME LENGTH (inches) = "; L0
     INPUT "ENTER RADOME BASE DIAMETER (inches) = "; D0

PRINT "": PRINT ""

   PRINT "ENTER INITIAL POINT WITHIN RADOME"
   PRINT ""
       INPUT "x0 ="; x0
       INPUT "y0 ="; y0
       INPUT "z0 ="; z0
```

```
PRINT "":   PRINT ""

PRINT "INPUT PROPAGATION VECTOR"
   INPUT "kx ="; kx
   INPUT "ky ="; ky
   INPUT "kz ="; kz

'    Begin Calculations Here:

' One Time Calculation for Radome Parameters R,B:

      B = (4 * L0 ^ 2 - D0 ^ 2) / (4 * D0)
      R = (4 * L0 ^ 2 + D0 ^ 2) / (4 * D0)

' ROOTFINDER TO FIND Ci FOR MONOTONICALLY DECREASING FUNCTION STARTS HERE:

'     First: find end points:

      Ci = 0
      xi = x0 + Ci * kx
      yi = y0 + Ci * ky
      zi = z0 + Ci * kz
      GOSUB 100
      F = ERF
      Ci = (L0 - z0) / kz
      xi = x0 + Ci * kx
      yi = y0 + Ci * ky
      zi = z0 + Ci * kz
      GOSUB 100
      G = ERF

'     Second: find initial roots:

      M = (G - F) / L0
      Ci(0) = -F / M
      Ci = Ci(0)
      xi = x0 + Ci * kx
      yi = y0 + Ci * ky
      zi = z0 + Ci * kz
      GOSUB 100
      ERF(0) = ERF
      IF ABS(ERF(0)) < .01 THEN GOTO 200
      F = F + ERF(0)
      Ci(1) = -F / M
      Ci = Ci(1)
      xi = x0 + Ci * kx
      yi = y0 + Ci * ky
      zi = z0 + Ci * kz
      GOSUB 100
      ERF(1) = ERF
      IF ABS(ERF(1)) < .01 THEN GOTO 200
```

```
'       Third: find new roots:

        FOR I = 1 TO 30 STEP 2
          M = (ERF(I) - ERF(I - 1)) / (Ci(I) - Ci(I - 1))
          F = ERF(I - 1) - M * Ci(I - 1)
          Ci(I + 1) = -F / M
          Ci = Ci(I + 1)
          xi = x0 + Ci * kx
          yi = y0 + Ci * ky
          zi = z0 + Ci * kz
          GOSUB 100
          ERF(I + 1) = ERF
          IF ABS(ERF(I + 1)) < .01 THEN GOTO 200
          F = F + ERF(I + 1)
          Ci(I + 2) = -F / M
          Ci = Ci(I + 2)
          xi = x0 + Ci * kx
          yi = y0 + Ci * ky
          zi = z0 + Ci * kz
          GOSUB 100
          ERF(I + 2) = ERF
          IF ABS(ERF(I + 2)) < .01 THEN GOTO 200
        NEXT I

' - - - - - - - - - - - - - - - - - - - - - - - - - - - - -
200  '  Continue

     '  Calculate Normal Vector at Ray Intercept Point:

        xp = xi: yp = yi: zp = zi
        rp = SQR(xp ^ 2 + yp ^ 2)
        nx = ((rp + B) / SQR((rp + B) ^ 2 + zp ^ 2)) * xp / rp
        ny = ((rp + B) / SQR((rp + B) ^ 2 + zp ^ 2)) * yp / rp
        nz = zp / SQR((rp + B) ^ 2 + zp ^ 2)

        PRINT ""
        PRINT ""

        PRINT "RAY INTERCEPT POINT IS: xp ="; xp; " yp ="; yp; " zp ="; zp
        PRINT "SURFACE NORMAL IS: nx ="; nx; " ny ="; ny; " nz ="; nz

        END

100  '  SUBROUTINE TANGENT OGIVE SHAPE

     '  to prevent overflows:
        IF xi = 0 AND yi = 0 THEN xi = 1E-12

        IF zi  <= L0 THEN
          ERF = (SQR(R ^ 2 - zi ^ 2) - B) - SQR(xi ^ 2 + yi ^ 2)
        END IF
        IF zi > L0 THEN
```

```
            ERF = -(SQR(R ^ 2 - (2 * L0 - zi) ^ 2) - B) -SQR(xi ^ 2 +
  yi ^ 2)
         END IF

RETURN
```

Appendix 7C: Program POLY Software Listing

Fifth order polynomial regression computer code ————————————

```
    REDIM A(10), MATRX(10, 10), SUMXP(10), SUMYXP(10)
    REDIM xp(200), rp(200)
```

Variable List

xp = axial distance from radome base (inches)
rp = radius to radome outer surface at distance zp from base (inches)
A0 ... A5 = polynomial coefficients computed by this code

```
{code}start:

    RESET

    COLOR 15, 1: CLS
    COLOR 0, 7
    LOCATE 2, 19
      PRINT " ________________________________________ "
    LOCATE 3, 19
      PRINT " _ 5th ORDER POLYNOMIAL CURVE FITTING _ "
    LOCATE 4, 19
      PRINT " ________________________________________ "
    COLOR 0, 1
    LOCATE 3, 59
      PRINT STRING$(2, 176)
    LOCATE 4, 59
      PRINT STRING$(2, 176)
    LOCATE 5, 21
      PRINT STRING$(40, 176)

'  Enter Data Here:

input.data:

    COLOR 23, 1
    LOCATE 25, 60: PRINT CHR$(34); "E"; CHR$(34); " to End Data Entry";
    COLOR 7, 1

    EOD$ = "N"
    N = 0
    ROW = 8
```

```
LOCATE 7, 35, 1, 0, 15: PRINT "xp-Input        rp-Output"
LOCATE 8, 35: PRINT "————          —————-"

N = 0

DO
  N = N + 1
  IF ROW < 24 THEN ROW = ROW + 1 ELSE ROW = 24

  LOCATE ROW, 20: PRINT ; "Point #"; LTRIM$(STR$(N));

  COLOR 15, 1
  LOCATE ROW, 38
  INPUT ; "", CH$
  IF UCASE$(CH$) = "E" THEN EXIT DO
  xp(N) = VAL(CH$)

  COLOR 7, 1
  LOCATE ROW, 37: PRINT xp(N);

  LOCATE ROW, 52
  INPUT ; "", CH$
  IF UCASE$(CH$) = "E" THEN EXIT DO
  rp(N) = VAL(CH$)

  LOCATE ROW, 51: PRINT rp(N)
  IF UCASE$(CH$) = "E" THEN EXIT DO
  rp(N) = VAL(CH$)
LOOP

N = N - 1
```

' Matrix Formulation:

```
CLS
PWR = 5
FOR ROW = 1 TO PWR + 1
  FOR COL = 1 TO PWR + 2
    XMAT(ROW, COL) = 0
    TRMAT(ROW, COL) = 0
    MATRX(ROW, COL) = 0
  NEXT COL
NEXT ROW

FOR ROW = 1 TO N
  FOR COL = 1 TO PWR + 1
    XMAT(ROW, COL) = xp(ROW) ^ (COL - 1)
  NEXT COL
NEXT ROW

FOR ROW = 1 TO PWR + 1
```

```
      FOR COL = 1 TO N
        TRMAT(ROW, COL) = xp(COL) ^ (ROW - 1)
      NEXT COL
    NEXT ROW

    FOR ROW = 1 TO PWR + 1
      FOR COL = 1 TO PWR + 1
        FOR k = 1 TO N
        MATRX(ROW, COL) = MATRX(ROW, COL) + TRMAT(ROW, k) * XMAT(k, COL)
        NEXT k
      NEXT COL
    NEXT ROW

    FOR ROW = 1 TO PWR + 1
      COL = PWR + 2
      FOR i = 1 TO N
        MATRX(ROW, COL) = MATRX(ROW, COL) + xp(i) ^ (ROW - 1) * rp(i)
      NEXT i
    NEXT ROW

'   Gauss Elimination:

    FOR COL = 1 TO PWR
      MAXROW = COL

      FOR ROW = COL + 1 TO PWR + 1
        IF ABS(MATRX(ROW, COL)) > ABS(MATRX(MAXROW, COL)) THEN MAXROW
= ROW
      NEXT ROW

      IF MAXROW <> COL THEN
        FOR J = COL TO PWR + 2
          TMP = MATRX(COL, J)
          MATRX(COL, J) = MATRX(MAXROW, J)
          MATRX(MAXROW, J) = TMP
        NEXT J
      END IF

      FOR ROW = COL + 1 TO PWR + 1
        FTR = MATRX(ROW, COL) / MATRX(COL, COL)
          FOR J = COL + 1 TO PWR + 2
            MATRX(ROW, J) = MATRX(ROW, J) - FTR * MATRX(COL, J)
          NEXT J
        NEXT ROW
      NEXT COL

'   Backward Substitution:

      A(PWR + 1) = MATRX(PWR + 1, PWR + 2) / MATRX(PWR + 1, PWR + 1)
      i = PWR

      DO
```

```
      ITMP = i + 1
      TOT = 0
      FOR J = ITMP TO PWR + 1
        TOT = TOT + MATRX(i, J) * A(J)
      NEXT J
      A(i) = (MATRX(i, PWR + 2) - TOT) / MATRX(i, i)
      i = i - 1
      IF i < 1 THEN EXIT DO
    LOOP

'  Display Coefficients:

    COLOR 7, 1
    CLS

    LOCATE 1, 1
    FOR i = 0 TO PWR
      PRINT "A"; LTRIM$(STR$(i)); " = "; A(i + 1);
      PRINT ""
    NEXT i

    LOCATE 2, 20

'  Test Polynomial:

    COLOR 23, 1
    LOCATE 25, 70
    PRINT CHR$(34); "Q"; CHR$(34); " to Quit";

    DO
      COLOR 15, 1
      LOCATE 9, 1: PRINT SPACE$(80);
      LOCATE 9, 1: INPUT "Enter Value of xp to Test Polynomial: ", xp$
      IF xp$ = "Q" OR xp$ = "q" THEN GOTO run.new.case
      xp = VAL(zp$)
      rp = 0

      FOR power = 0 TO 5
        rp = rp + A(power + 1) * xp ^ power
      NEXT power

      COLOR 7, 1
      LOCATE 12, 1: PRINT SPACE$(80)
      COLOR 15, 0
      LOCATE 12, 22: PRINT " Result:  xp = "; xp; "   rp = "; rp
    LOOP

'  New Interpolation or End:

run.new.case:
    CLS
    LOCATE 13, 19: INPUT "Run Interpolation with New Data (Y/N)"; A$
```

```
IF UCASE$(A$) = "Y" THEN
  ERASE SUMXP, MATRX, SUMYXP, A
  REDIM A(6), MATRX(6, 7), SUMXP(10), SUMYXP(6)
  GOTO start
ELSE
  COLOR 11, 0
  CLS
  END
END IF
```

Appendix 7D: Program COMPOSE Software Listing

Code to demonstrate decomposition and recomposition of electromagnetic wave components when propagating through a radome wall

```
REDIM k(3), n(3), d(3), Ei(3), PSi(3), Et(3), PSt(3), REt(3), IEt(3)
```

Note that indices 1,2,3 in above variables all correspond to x,y,z coordinates, respectively.

```
PI = 3.14159265#: RAD = PI / 180
```

Variable List

k(1),k(2),k(3)	= x,y and z components of propagation vector k, respectively (inches)
n(1),n(2),n(3)	= x,y and z components of surface normal vector n, respectively(inches)
d(1),d(2),d(3)	= a vector normal to the plane formed by vectors n and k. i.e., d = n x k, and is a parallel reference vector (inches)
g(1),g(2),g(3)	= a vector parallel to the plane formed by n and k. i.e., g = d x k, and is a perpendicular reference vector (inches)
Ei(1),Ei(2),Ei(3)	= x,y and z components of the incident electromagnetic field (v/m)
PSi(1),PSi(2),PSi(3)	= x,y and z components of the phase of the incident electromagnetic field (radians)
Tpara,Tperp	= parallel and perpendicular polarization voltage transmission coefficients, respectively(magnitude - a dimensionless quantity)
IPDpara,IPDperp	= parallel and perpendicular polarization insertion phase delays (IPD's) associated with the voltage transmission coefficients(radians)

Et(1),Et(2),Et(3) = x,y and z components of the transmitted electromagnetic field (v/m)
PSt(1),PSt(2),PSt(3) = x,y and z components of the phase of the transmitted electromagnetic field (radians)

```
Enter Data Here:
   COLOR 15, 1
   CLS
   PRINT "INPUT PROPAGATION VECTOR"
   INPUT "kx ="; k(1)
   INPUT "ky ="; k(2)
   INPUT "kz ="; k(3)
   PRINT "":  PRINT ""

   PRINT "INPUT MAGNITUDE OF INCIDENT ELECTROMAGNETIC FIELD"
   INPUT "Eix ="; Ei(1)
   INPUT "Eiy ="; Ei(2)
   INPUT "Eiz ="; Ei(3)
   PRINT "":   PRINT ""

   PRINT "INPUT PHASE OF INCIDENT ELECTROMAGNETIC FIELD"
   INPUT "PSix ="; PSi(1)
   INPUT "PSiy ="; PSi(2)
   INPUT "PSiz ="; PSi(3)
   PRINT "":   PRINT ""

   PRINT "INPUT SURFACE NORMAL VECTOR"
   INPUT "nx ="; n(1)
   INPUT "ny ="; n(2)
   INPUT "nz ="; n(3)
   PRINT "":   PRINT ""

   PRINT "INPUT MAGNITUDE OF VOLTAGE TRANSMISSION COEFFICIENTS"
   INPUT "Tpara ="; Tpara
   INPUT "Tperp ="; Tperp
   PRINT "":   PRINT ""

   PRINT "INPUT PHASE OF THE VOLTAGE TRANSMISSION COEFFICIENTS"
   INPUT "IPDpara ="; IPDpara
   INPUT "IPDperp ="; IPDperp
   PRINT "":   PRINT ""

'  Program Computations Begin Here:
'  Calculate Perpendicular Unit Vector, D :
mag = SQR((n(2) * k(3) - n(3) * k(2)) ^ 2 + (n(3) * k(1) - n(1) * k(3)) ^
 2 + (n(1) * k(2) - n(2) * k(1)) ^ 2)
       d(1) = (n(2) * k(3) - n(3) * k(2)) / mag
       d(2) = (n(3) * k(1) - n(1) * k(3)) / mag
       d(3) = (n(1) * k(2) - n(2) * k(1)) / mag

'  Calculate Parallel Unit Vector, G :
```

```
    mag = SQR((d(2) * k(3) - d(3) * k(2)) ^ 2 + (d(3) * k(1) -d(1) *
k(3)) ^ 2 + (d(1) * k(2) - d(2) * k(1)) ^ 2)
       g(1) = (d(2) * k(3) - d(3) * k(2)) / mag
       g(2) = (d(3) * k(1) - d(1) * k(3)) / mag
       g(3) = (d(1) * k(2) - d(2) * k(3)) / mag

'  Project Incident Wave Thru Wall:
       REtpara = 0: IEtpara = 0: REtperp = 0: IEtperp = 0

     FOR I = 1 TO 3
       REtpara = REtpara + Ei(I) * g(I) * Tpara * COS(PSi(I) + IPDpara)
       IEtpara = IEtpara + Ei(I) * g(I) * Tpara * SIN(PSi(I) + IPDpara)

       REtperp = REtperp + Ei(I) * d(I) * Tperp * COS(PSi(I) + IPDperp)
       IEtperp = IEtperp + Ei(I) * d(I) * Tperp * SIN(PSi(I) + IPDperp)
     NEXT I

     FOR I = 1 TO 3
       REt(I) = REtpara * g(I) + REtperp * d(I)
       IEt(I) = IEtpara * g(I) + IEtperp * d(I)
          IF REt(I) = 0 AND IEt(I) = 0 THEN
          Pst(I) = 0: Et(I) = 0
               GOTO 100
          END IF
       Et(I) = SQR(REt(I) ^ 2 + IEt(I) ^ 2)
       PSt(I) = ATN(ABS(IEt(I) / REt(I)))

100  '  CONTINUE
     NEXT I

'  END OF COMPUTATIONS:
PRINT "Eix ="; Ei(1); "PSix ="; PSi(1); "Eiy ="; Ei(2); "PSiy ="; PSi(2)
PRINT "Etx ="; Et(1); "PStx ="; PSt(1); "Ety ="; Et(2); "PSty ="; PSt(2)
END
```

Chapter 8

Radome-Enclosed Monopulse Antennas

This chapter introduces computerized techniques to analyze radome-enclosed monopulse antenna systems. A monopulse antenna functions to detect target location. Typically, radomes associated with these antennas are aerodynamically streamlined and have a high fineness ratio (up to three or greater). For a specific application, the radome materials must be chosen to simultaneously satisfy structural, environmental, and electrical requirements. The computerized analysis techniques in this chapter can determine the effect of the radome on the performance of the enclosed antenna and its electronics processing function.

A monopulse antenna is generally mounted forward in an aircraft or missile, as depicted in Figure 8.1. The term *monopulse* refers to a signal-processing technique in which only a single radar return pulse is needed to resolve a target's location in both the elevation and azimuth planes of the antenna. Monopulse systems are typically active radar, but in certain applications they may be passive receptors. In this book, the term *seeker* is used synonymously with monopulse electronics systems.

Ideally, a radome protects the monopulse antenna from the adverse environment while having an insignificant effect on the electrical performance of the enclosed antenna. However, in practice, a radome always changes the electrical performance of the monopulse antenna because of the following factors: (1) dissipative losses within a dielectric material, (2) electrical phase shifts introduced by the presence of the radome, and (3) internal reflections. These changes produced by the radome manifest themselves as follows:

- Decreased antenna gain;
- Increased antenna beamwidth and sidelobe levels;
- Decreased monopulse null depth;
- Depolarization of the wave propagated through the radome walls;
- Introduction of boresight error (BSE) and boresight error slope (BSES), which are functions of the antenna scan angle [1,2].

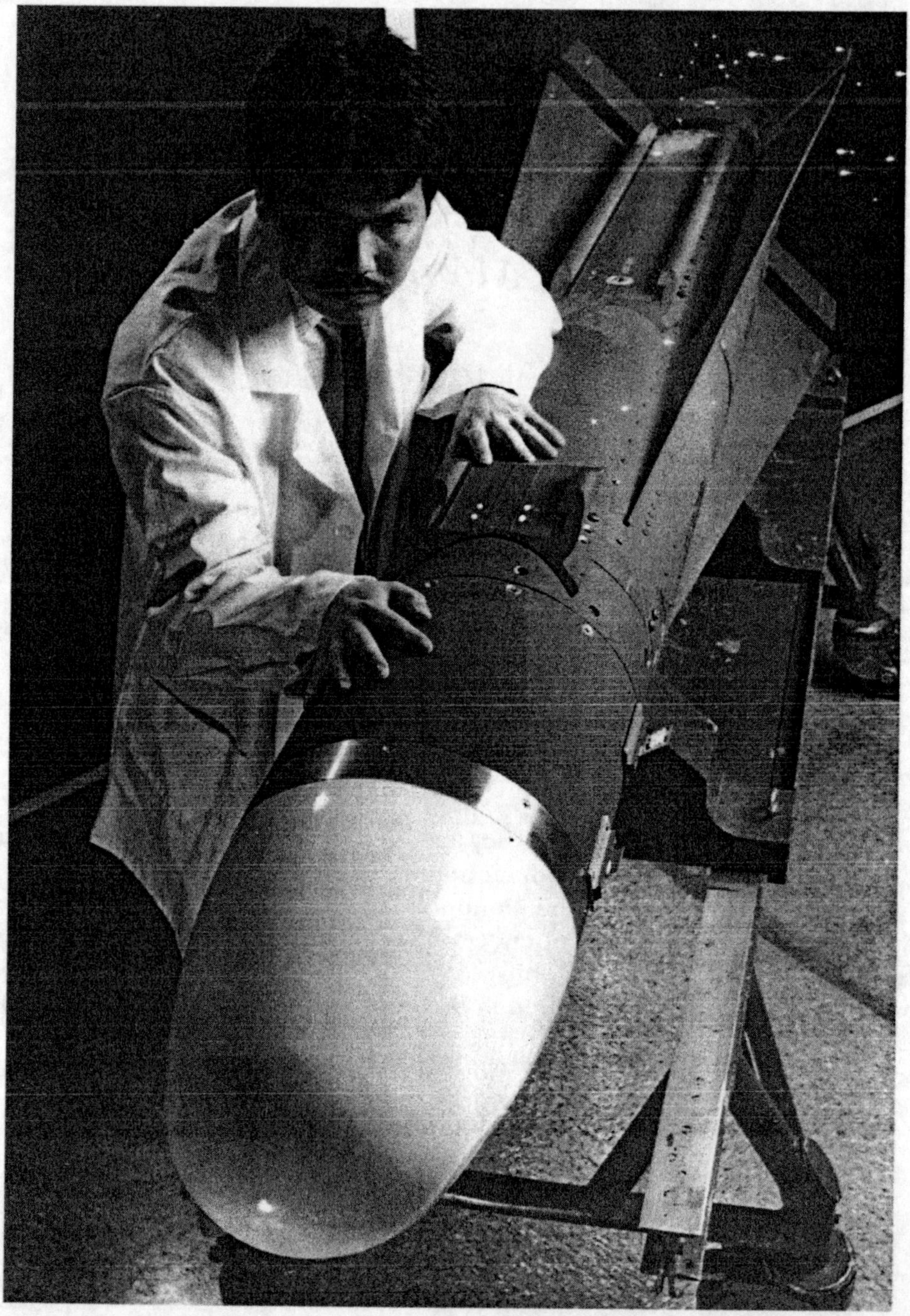

Figure 8.1 Radome-enclosed monopulse antenna used for missile guidance. (Photograph courtesy of Hughes Aircraft Missile Systems Division.)

In this chapter, antenna pattern calculations are made in steps, which must be performed in sequence as follows:

1. Transpose all aperture points to radome coordinates.
2. Transform all ray vectors from aperture sample points in the pattern look direction
3. Compute the antenna space phase terms.
4. Determine all ray intercept points on the radome surface.
5. Determine angles-of-incidence of these rays with respect to the radome surface normal at each intercept point.
6. Compute complex voltage transmission coefficients for each ray.
7. Perform an antenna aperture integration

Section 8.1 provides a definition of BSE and BSES. Section 8.2 discusses antenna patterns and monopulse error voltages for an antenna without a radome, whereas Section 8.3 deals with an antenna enclosed within a radome. Section 8.4 covers radome modeling considerations, including rain erosion, aerothermal heating, and radome effects on a conscan antenna.

8.1 DEFINITION OF BORESIGHT ERROR AND BORESIGHT ERROR SLOPE

When used in missiles, the seeker antenna is a key part of the guidance system, which also includes the radome, receiver, autopilot, and airframe. The seeker has several functions:

- To measure the line-of-sight (or line-of-sight angular rates);
- To measure the range-to-target and the closing velocity;
- To provide steering commands to the missile autopilot and control surfaces [3–7].

Figure 8.2 shows a representative closed-loop servo control system, in which the signals received by the missile-mounted seeker antenna are used to guide the missile to the target. This guidance is accomplished in the tracking loop, which is driven by the so-called monopulse elevation and azimuth plane guidance error voltages.

The seeker initially finds the target in angle space through the seeker antenna's beamwidth. The antenna beam is steered by one of the following methods:

- A gimbal that mechanically moves the entire antenna structure;
- An electromechanical mechanism that moves only a portion of the antenna, such as a subreflector;
- An electronic steering mechanism, such as a phased-array antenna;
- Movement of the entire airframe, in the case of an antenna that is stationary

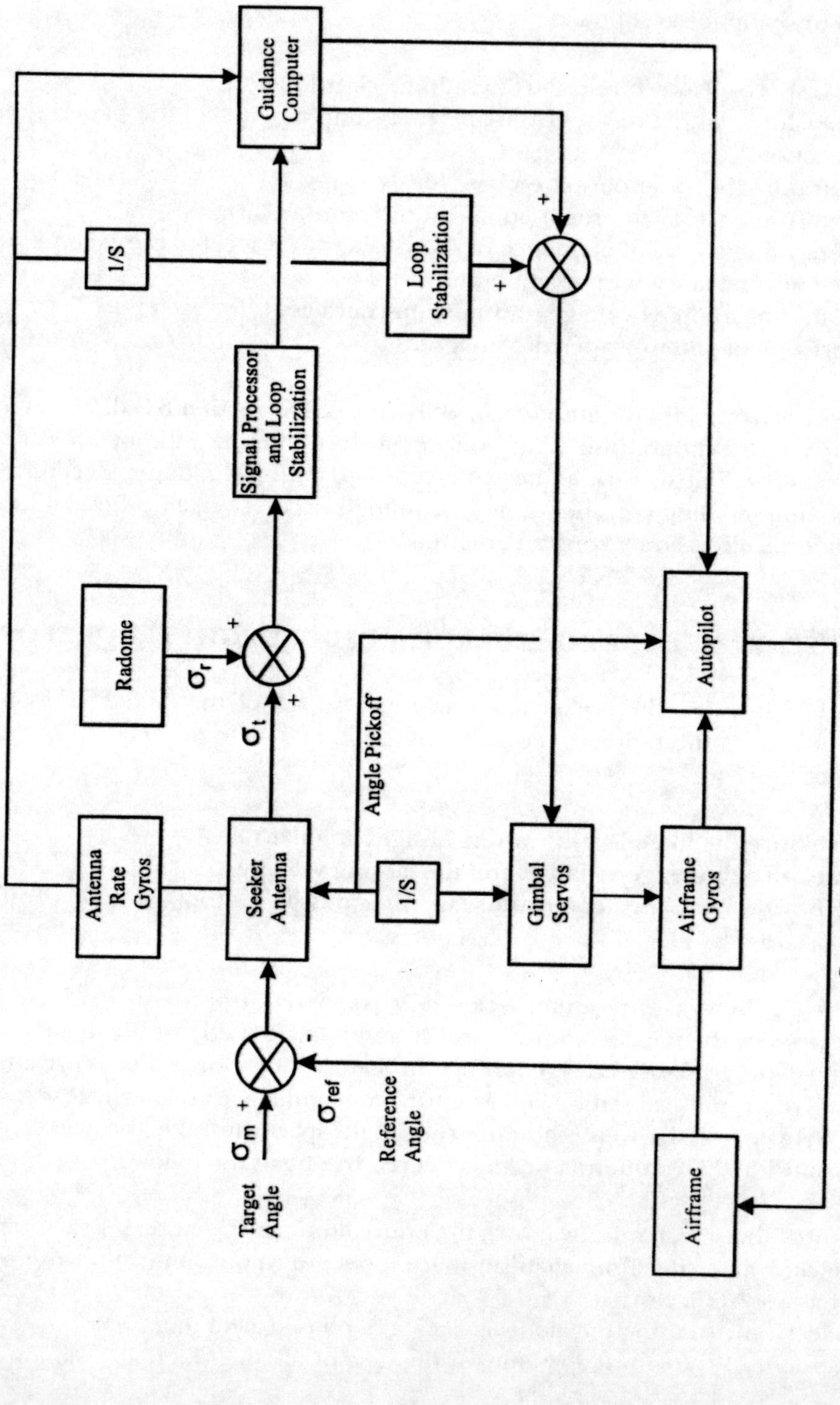

Figure 8.2 Representative guidance system block diagram.

and forward-looking. Fixed seeker antenna beams are used only when a pursuit navigation guidance algorithm is used. Movable seeker antenna beams are used with proportional navigation (or modern guidance law) guidance algorithms [8].

Figure 8.3 illustrates the definitions of the angles relating to the seeker-radome and missile-target geometry. In this figure, the seeker antenna is mounted on a mechanical gimbal whose pointing function is controlled by a servo loop. Because of the diffraction errors introduced by the presence of the radome, the servo system will try to keep the antenna pointed toward the apparent target direction, as opposed to the actual target direction.

The difference between the actual missile-to-target direction and the apparent missile-to-target direction is known as the *radome boresight error* (BSE). Conceptually, BSE is an angular shift in the apparent target direction and generally corresponds to difference pattern minimums in the far-field radiation patterns of the radome-enclosed guidance antenna. The *boresight error slope* (BSES) is the rate of change of the boresight error with respect to antenna scan angle.

The reference angle shown in Figure 8.3 is derived from airframe gyros. The antenna pointing angles are often determined from rate gyros mounted on the antenna gimbal. The monopulse error voltages are fed back to the autopilot. In response to the autopilot commands, the missile changes attitude, which causes the antenna to receive through an ever-changing portion of the radome. As a consequence, the response characteristics of the airframe and autopilot generate a pseudo-noise influence due to these radome-induced errors [9].

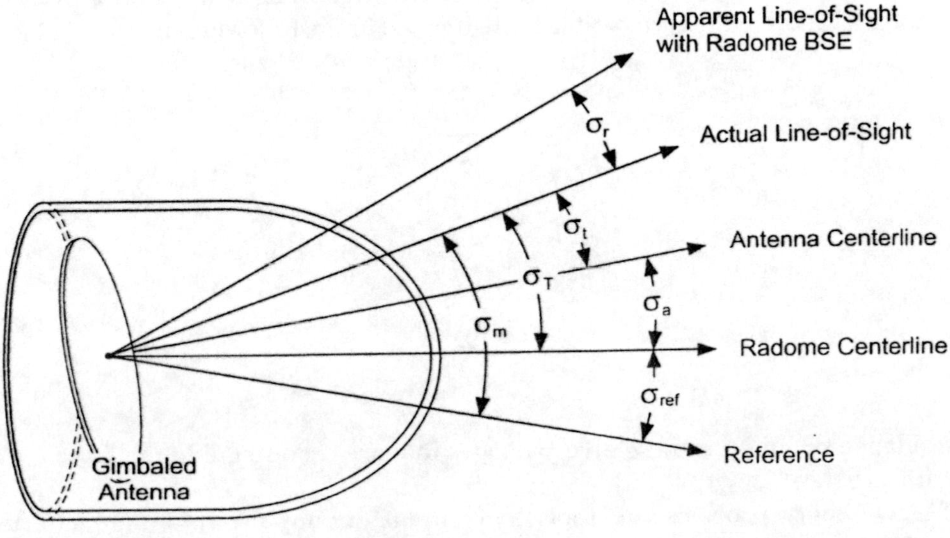

Figure 8.3 Monopulse antenna-to-target angular relationships.

8.2 CALCULATION OF ANTENNA PATTERNS AND MONOPULSE ERROR VOLTAGES FOR AN ANTENNA WITHOUT A RADOME

The physical aperture of a monopulse antenna is usually divided into four quadrants. Each of these is fed by a microwave network that resolves the signals received by each of the four quadrants into a sum channel, and into elevation and azimuth plane difference channels, as illustrated in Figure 8.4. A monopulse radar receiver processes these sum and difference channel signals as complex voltages, which are resolved into in-phase (I) and out-of-phase or phase quadrature (Q) voltages prior to processing by the signal processor.

The sum channel antenna pattern corresponds to a single on-boresight antenna beam, with the half-power beamwidth dictated by both the antenna aperture size in wavelengths and the aperture illumination function. The elevation channel antenna pattern has a boresight null only in the elevation plane, whereas the azimuth channel antenna pattern has a boresight null only in the azimuth plane. The effective sum channel beamwidth obtained with a monopulse antenna is the beamwidth of the aperture, and the maximum gain of the aperture is realized.

Early monopulse systems used reflector antennas or lens antennas, with the monopulse-beam-forming network achieved with a large, complex waveguide monopulse network. Modern approaches to achieve sum and difference channel beams have very small beam-forming networks, with little or no aperture blockage [10]. These approaches include waveguide flat-plate-array and microstrip-array configurations. The photograph of Figure 8.5 shows a microstrip, monopulse array antenna, utilizing patch radiators and having the feed network on the same substrate as the patch radiators.

From the monopulse sum and difference channel voltages, the monopulse azimuth and elevation plane error voltages are derived in the following manner [11]:

$$V_{az} = \frac{\Delta_{az}}{\Sigma} \tag{8.1}$$

$$V_{el} = \frac{\Delta_{el}}{\Sigma} \tag{8.2}$$

In a guidance system, it is these error voltages that are used to guide the missile toward the target.

Array theory provides the tools to compute monopulse antenna patterns without the need to apply exact electromagnetic modeling. The following analysis introduces the modeling methodology, and we use the spherical coordinate system

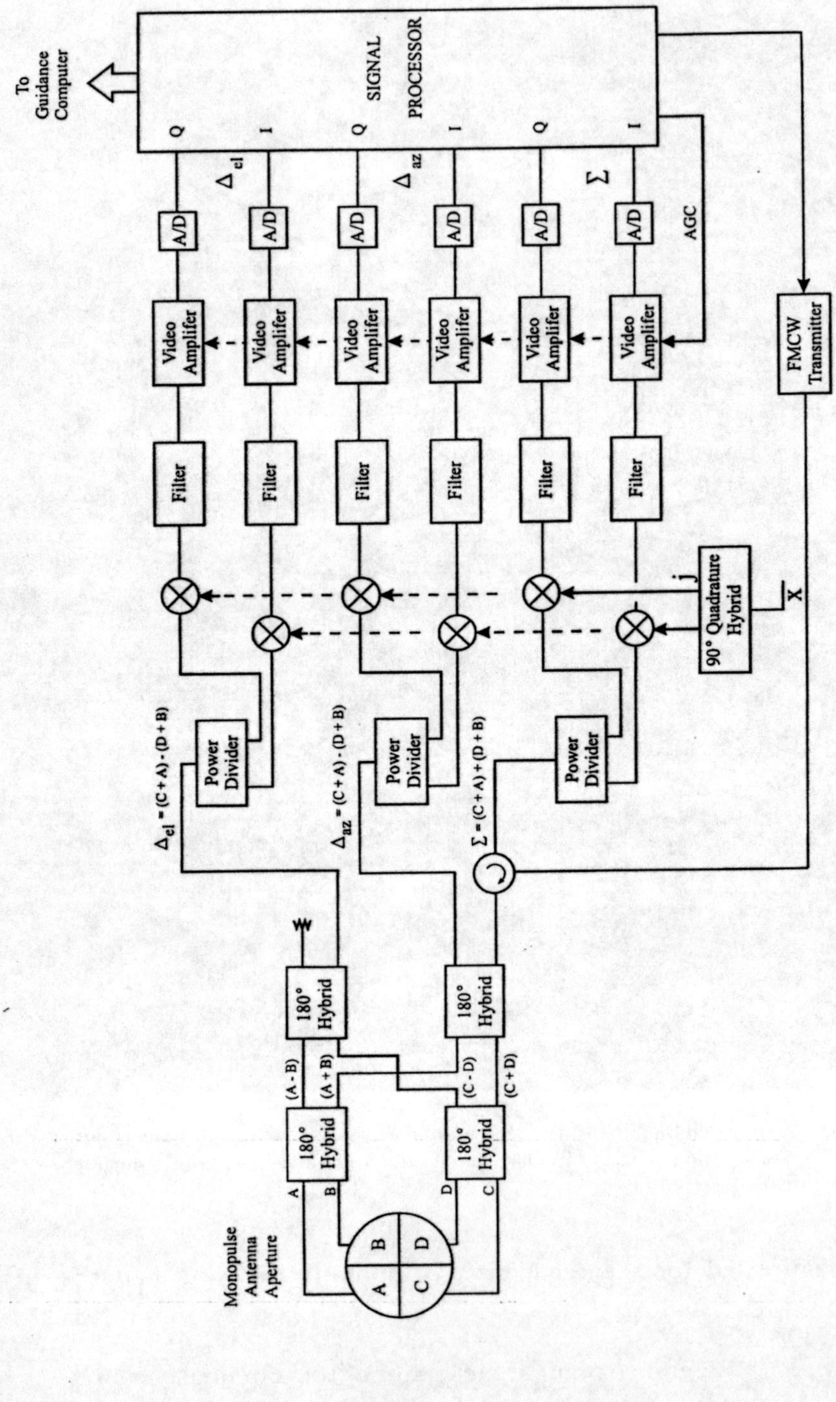

Figure 8.4 Block diagram of a typical monopulse radar.

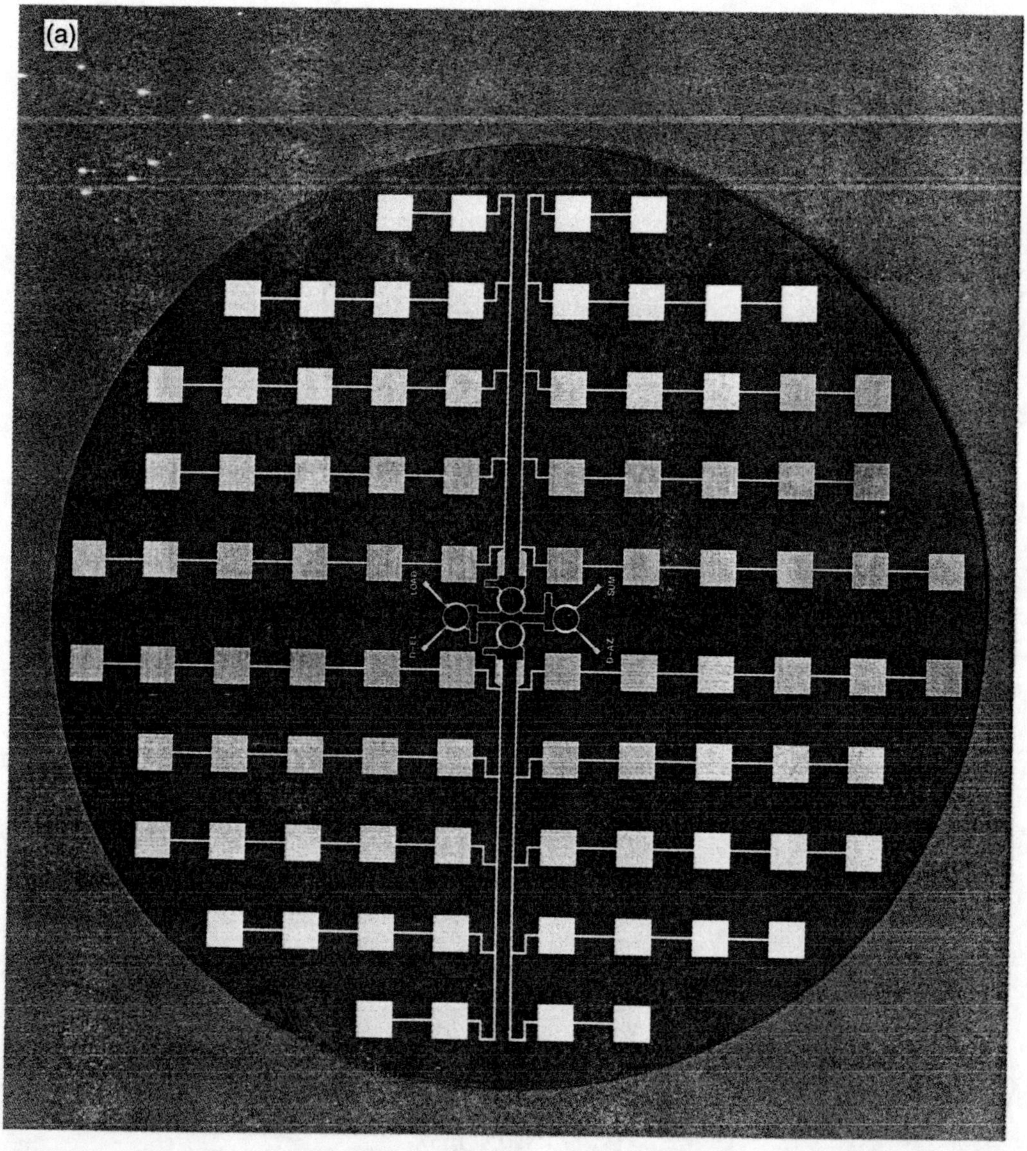

Figure 8.5 (a) Square patch radiator array with monopulse comparator on substrate. (Photo courtesy of Ball Corporation.) (b) Circular patch radiator array. (Photo courtesy of Seavey Engineering.)

shown in Figure 8.6 for the calculation of antenna radiation patterns. In this system:

$P(x_o, y_o, z_o)$ = Reference point at the center of the antenna aperture;
$P(x, y, z)$ = Observation point at distance R from the antenna;

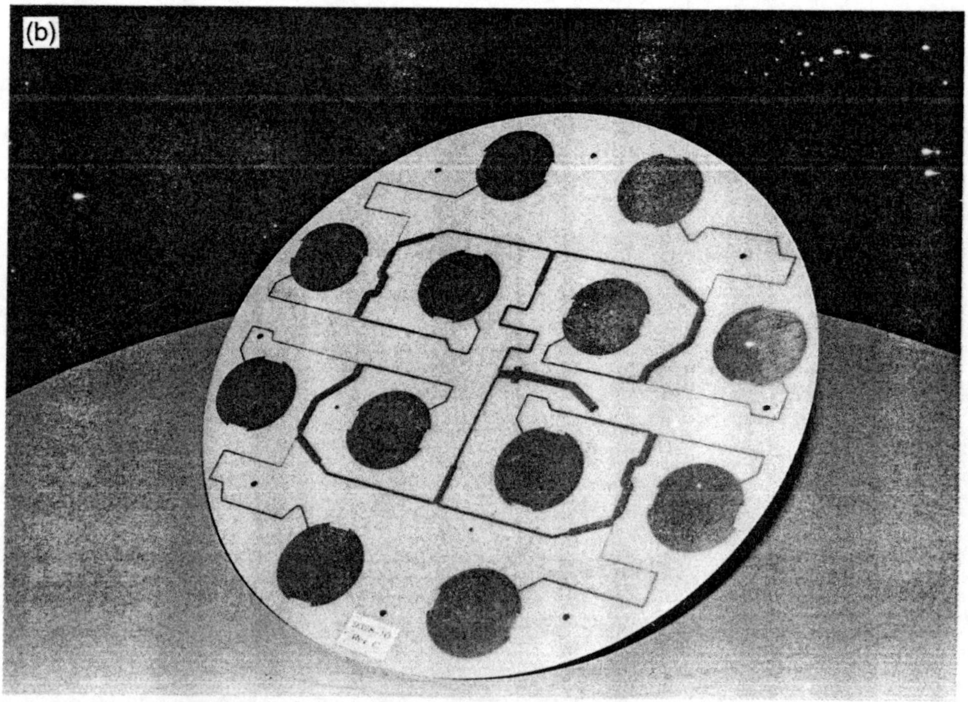

Figure 8.5 (continued)

$P(x_i, y_i, z_i)$ = Location of i'th sample point on the antenna aperture;
r_o = Unit vector in direction from $P(x_o, y_o, z_o)$ to $P(x, y, z)$;
r_i = Vector in direction from $P(x_o, y_o, z_o)$ to $P(x_i, y_i, z_i)$

The circular antenna aperture lies entirely in the x-y plane and has a radius given by the formula

$$r_a = \sqrt{(x_i - x_o)^2 + (y_i - y_o)^2 + (z_i - z_o)^2} = \frac{D_a}{2} \tag{8.3}$$

where D_a is the antenna diameter.

It is customary to obtain current sample points at spacings d_x and d_y in the x and y planes, respectively. If these spacings are a half-wave, or less, at the highest operating frequency, we guarantee that there will be no aliasing into real space to cause computational errors in the antenna pattern prediction. However, if we are only interested in the mainlobe and first sidelobes of the radiation patterns, informal testing by the author shows that sample spacings up to a wavelength or more may be acceptable with only small computational errors.

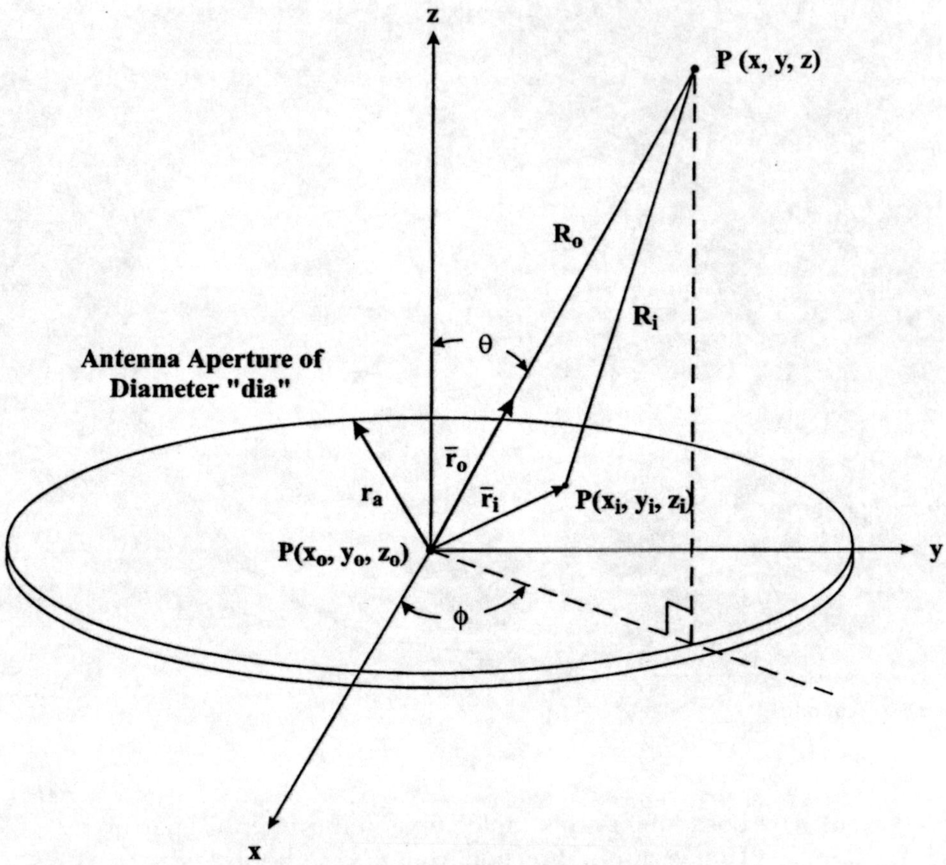

Figure 8.6 Coordinate system used in antenna pattern calculations.

In accordance with Appendix A at the end of the book, we will now assume that the sample point currents are an ensemble of infinitesimal dipoles, each of which is weighted in both amplitude and phase. The method of determining the space phase at the i'th sample point is also illustrated in Figure 8.6.

Points on the surface of the antenna aperture are uniquely defined via integers m,n:

$$r_i = md_x\bar{x} + nd_y\bar{y} \tag{8.4}$$

For distances relatively far from the i'th infinitesimal dipole, its radiation pattern has an exponential phase dependance as it expands as a spherical wave, and also has a sine-weighted amplitude dependance with a radiation maximum normal to the dipole. Based on the coordinate system selected, if the dipoles are y-directed, an

appropriate coordinate transformation derives the radiation element pattern for the i'th infinitesimal dipole as the following:

$$E_i = \cos[\tan^{-1}(\tan\theta \sin\phi)] \frac{e^{-jkR_i}}{R_i} \tag{8.5}$$

In this equation, k is the free-space wave number at the frequency of interest. If the dipoles are x-directed, the radiation element pattern becomes:

$$E_i = \cos[\tan^{-1}(\tan\theta \cos\phi)] \frac{e^{-jkR_i}}{R_i} \tag{8.6}$$

Under two conditions, the cosine factors in (8.5) and (8.6) will be near unity over the angle space corresponding to the antenna main beam through the first sidelobes. These two conditions are (1) the antenna diameter is larger than about five wavelengths and (2) we are only interested in the computation of a limited portion of the antenna patterns, such as the main lobe and the first sidelobes.

If the cosine factors are ignored, the impact on computational accuracy will be small when calculating radome transmission loss and first sidelobe degradation. However, when estimating far-out sidelobes, the cosine factors are important.

Accordingly, based on the MN total current sample points, the antenna array pattern can be determined from the superposition of the infinitesimal dipoles over the entire aperture:

$$E_T = \sum_{m=1}^{M} \sum_{n=1}^{N} F_{mn}^{a} \frac{e^{-ik\overline{R}_{mn}}}{R_{mn}} \tag{8.7}$$

Here, the coefficients F_{mn}^{a} are the applied current element weight at the sample points induced by a received signal. In general, these weighting coefficients are complex quantities.

In (8.7), the required computational distance R_{mn} should be sufficiently into the far field of the antenna. The actual value of R_{mn} depends on the degree of fine structure desired in the antenna pattern computation. The conventional definition of antenna far-field corresponds to the condition wherein

$$R_{mn} = \frac{2D_a^2}{\lambda} \tag{8.8}$$

A one far-field distance is adequate for mainlobe and first sidelobe antenna pattern calculations. However, for antenna aperture distributions that give very low sidelobe antenna patterns, it may be necessary to use distances at least five times greater than those of (8.8) [12,13]. If we let the computational distance approach infinity, the following approximation holds:

$$\frac{e^{-jkR_{mn}}}{R_{mn}} \approx \left[\frac{e^{-jkR_o}}{R_o}\right] e^{jk\bar{r}_{mn}\cdot\bar{r}_o} \tag{8.9}$$

Under this condition,

$$E_T = \left[\frac{e^{-jkR_o}}{R_o}\right] \sum_{m=1}^{M} \sum_{n=1}^{N} F^a_{mn}\, e^{jk\bar{r}_{mn}\cdot\bar{r}_o} \tag{8.10}$$

It is customary to drop out the left factor in brackets, [ ], because the antenna pattern is generally described on a sphere of a constant radius, and this term just becomes a normalizing constant [14]. This factor will have no impact on the computation of radome parameters of interest. The antenna pattern then takes the form:

$$E_T = \sum_{m=1}^{M} \sum_{n=1}^{N} F^a_{mn}\, e^{jk\bar{r}_{mn}\cdot\bar{r}_o} \tag{8.11}$$

Using vector mathematics, we note that

$$\bar{r}_{mn} = (x_{mn} - x_o)\bar{x} + (y_{mn} - y_o)\bar{y} + (z_{mn} - z_o)\bar{z} \tag{8.12}$$

Also, the unit vector in the direction of the observation point is given in terms of the directional cosines [15]:

$$\bar{r}_o = \sin\theta\cos\phi\bar{x} + \sin\theta\sin\phi\bar{y} + \cos\theta\bar{z} \tag{8.13}$$

From these last two equations, we can simplify the dot product in (8.10):

$$\bar{r}_{mn}\cdot\bar{r}_o = (x_{mn} - x_o)\sin\theta\cos\phi + (y_{mn} - y_o)\sin\theta\sin\phi + (z_{mn} - z_o)\cos\theta \tag{8.14}$$

If the center of the antenna is at the point $x_o = y_o = z_o = 0$ and we select all $z_{mn} = 0$, then the simplification on the array pattern of the antenna takes the form:

$$E_T = \sum_{m=1}^{M} \sum_{n=1}^{N} F_{mn}^{a} e^{-jk(x_{mn} \sin\theta\cos\phi + y_{mn} \sin\theta\sin\phi)} \tag{8.15}$$

or

$$E_T = \sum_{m=1}^{M} \sum_{n=1}^{N} F_{mn}^{a} e^{-j(k_x x_{mn} + k_y y_{mn})} \tag{8.16}$$

where k_x and k_y are the propagation constants in the x and y planes, respectively.

The monopulse sum and difference patterns should be computed by sequentially loading this equation with the proper aperture distribution functions. For example, for a monopulse antenna, these weighting factors may correspond to the following theoretical distributions:

$$F_{mn}^{a} = F_{\Sigma}^{a} = \cos\left(\frac{\pi}{2} \frac{r_{mn}}{r_a}\right) \tag{8.17}$$

$$F_{mn}^{a} = F_{\Delta el}^{a} = \sin\left(\pi \frac{y_{mn}}{r_a}\right) \tag{8.18}$$

or

$$F_{mn}^{a} = F_{\Delta az}^{a} = \sin\left(\pi \frac{x_{mn}}{r_a}\right) \tag{8.19}$$

If our computation is only interested in the relative directivity normalized with respect to the peak of the sum pattern, then the antenna array pattern is

$$E_T = \frac{\sum_{m=1}^{M} \sum_{n=1}^{N} F_{mn}^{a} e^{-jk(x_{mn} \sin\theta\cos\phi + y_{mn} \sin\theta\sin\phi)}}{\sum_{m=1}^{M} \sum_{n=1}^{N} \cos\left(\frac{\pi}{2} \frac{r_{mn}}{r_a}\right)} \tag{8.20}$$

where the F_{mn}^{a} in the numerator is one of the appropriate weighting factors from (8.17) through (8.19). The peak of the relative directivity sum pattern will then correspond to unity or 0 dB.

Figure 8.7 shows the software flow diagram for a computer program designed to compute antenna patterns for a monopulse antenna. For angles of interest in antenna angle space, θ and ϕ, a surface integration over the entire antenna aperture

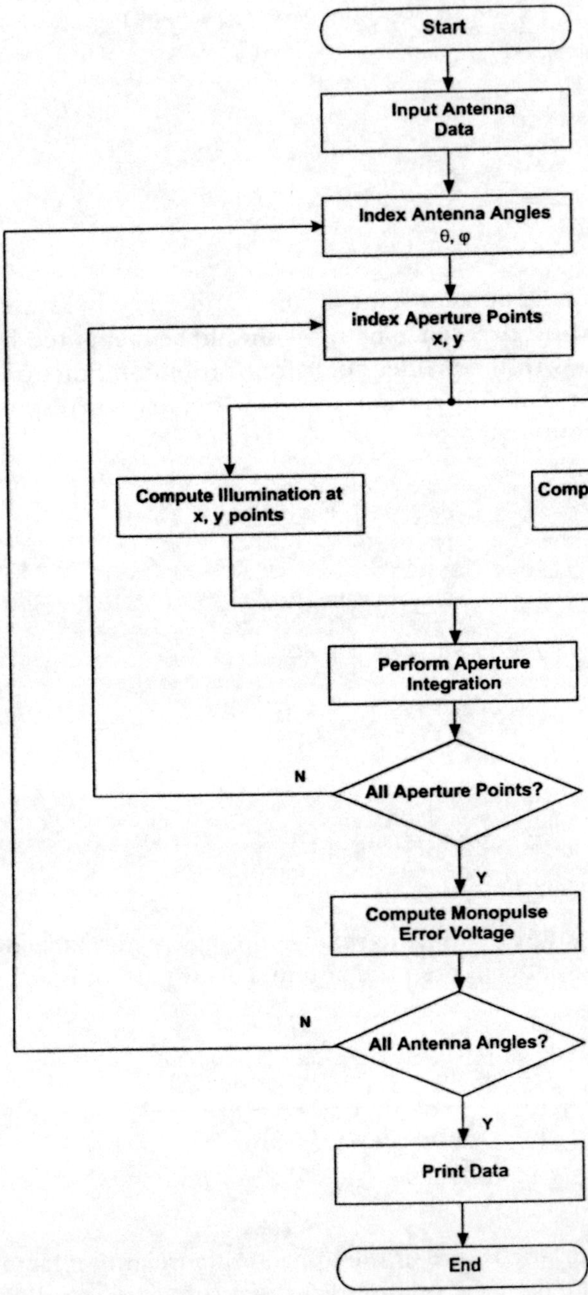

Figure 8.7 Computer program flow diagram for Program PATTERN.

must be performed for each of these three aperture weighting distributions. This integration will allow us to compute the antenna sum voltage and the two difference channel voltages.

Based on the flow diagram in Figure 8.7, a BASIC program was developed. Appendix 8A at the end of this chapter gives a complete software listing for the program. The software listing clearly defines all the input and output variables, which makes it user friendly. As configured, this code computes the principal plane sum and difference antenna pattern voltages, and also computes the monopulse error channel voltages.

To demonstrate use of the program, we assumed a 15-in. circular antenna aperture at 10 GHz, having the aperture distributions of (8.17) through (8.1). Figure 8.8 shows the computed E-plane and H-plane sum patterns. Figure 8.9 depicts the corresponding azimuth and elevation plane monopulse error channel voltages. Because there are no radome-induced errors, both of these functions are zero on the antenna boresight. In performing this modeling, we investigated the computational error in first sidelobe and antenna sum channel half-power beamwidth predictions versus the number of aperture sample points per wavelength. The resulting data are plotted in Figure 8.10.

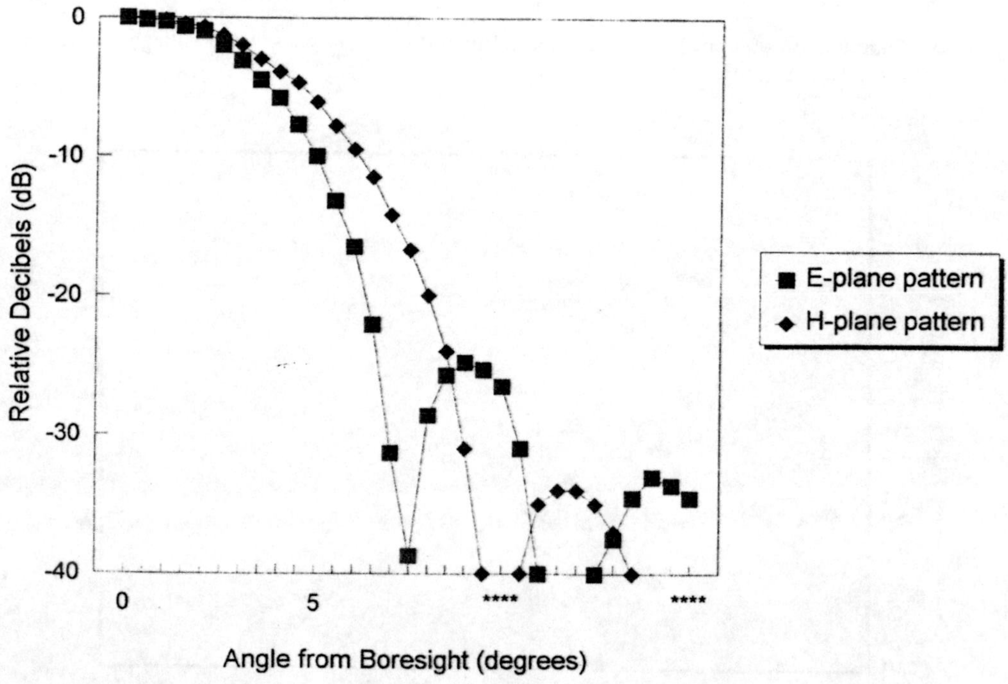

Figure 8.8 Computation of azimuth and elevation plane monopulse antenna sum pattern for a 15-in. aperture at 10 GHz.

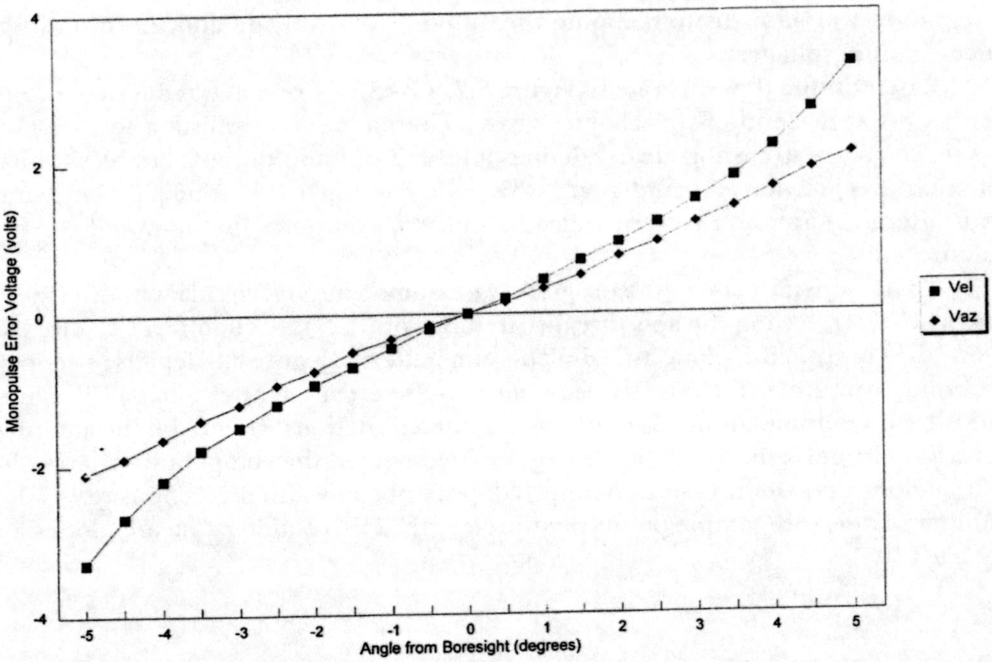

Figure 8.9 Computation of monopulse antenna azimuth and elevation plane error voltages for a 15-in. aperture at 10 GHz.

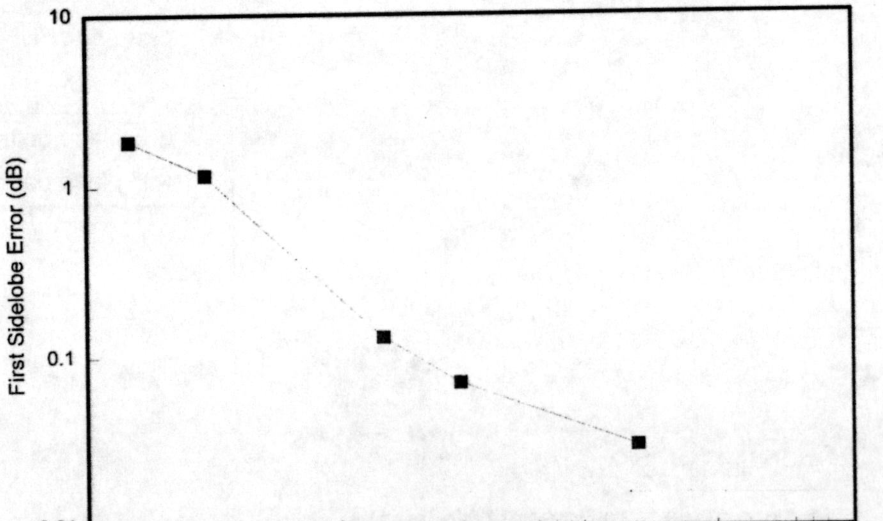

Figure 8.10 Computational error in antenna beamwidth and sidelobe versus sample granularity.

8.3 CALCULATION OF ANTENNA PATTERNS AND MONOPULSE ERROR VOLTAGES FOR A RADOME-ENCLOSED ANTENNA

Two approaches to boresight error computation are used: (1) Determine the shift in the difference patterns' null position, which results from wavefront phase distortion of the signal as it propagates through the radome [16] or (2) determine the far-field sum pattern peak location [17].

This chapter uses the first method, which derives boresight error in the same way that a monopulse seeker does. This method is based on the determination of the shift in the zero crossings of the monopulse error in channel voltages since the resulting wave diffraction with a radome causes a shift in the apparent direction of the target.

Previously in (8.1) and (8.2), we showed how monopulse azimuth and elevation plane error voltages are derived from the monopulse antenna sum and difference port voltages. When no radome is present, the zero crossings in these monopulse error voltage channels correspond to the condition of a target on the antenna axis, that is, normal to the surface of the antenna. With a radome over the antenna, as the electromagnetic signal propagates through the dielectric radome wall, the resulting wave diffraction causes a shift in the apparent direction of the target.

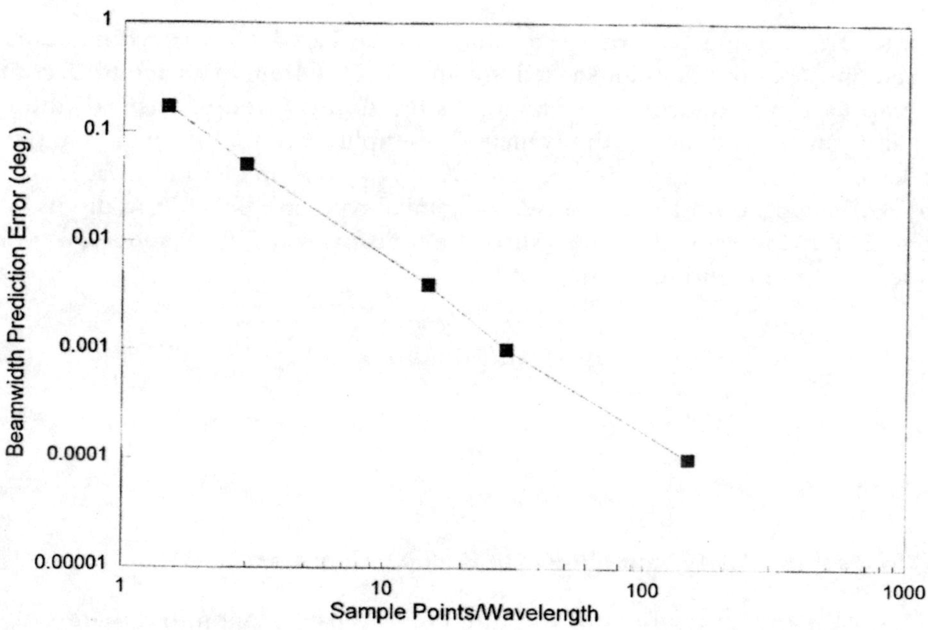

Figure 8.10 (continued)

With the antenna physically boresighted on the target, the azimuth and elevation boresight errors are computed from the following formulas [18]:

$$\mathrm{BSE}_{az} = \frac{V_{az}}{K_{az}} \tag{8.21}$$

$$\mathrm{BSE}_{el} = \frac{V_{el}}{K_{el}} \tag{8.22}$$

In these equations, the constants K_{az} and K_{el} are the azimuth and elevation difference channel monopulse antenna sensitivity constants, respectively, computed (or measured) for the no-radome condition. Usually, these constants have units of volts-per-milliradian (mrad), and the boresight error has units of milliradians. BSE is highly influenced by antenna polarization and its implication for the azimuth and elevation plane error voltages.

The following steps summarize the modeling process for obtaining a computerized solution to a radome-enclosed monopulse antenna. In these steps, the antenna is assumed to be in a receiving mode.

Step 1: Compute the Antenna Pattern Space Phase Terms

Compute the antenna pattern space phase terms, based on a coordinate system centered on the antenna as illustrated in Figure 8.11. Often, in an actual aircraft or missile application, description of a target's location is given in terms of the roll, pitch, and yaw coordinates of the vehicle. To simplify our analysis, a gimbal system was selected for modeling. It describes a target's position in elevation (EL) and azimuth (AZ) angles. Both EL/AZ or AZ/EL gimbal types are possible, as discussed in Chapter 2. The following equations show the transformation from spherical coordinates selected to EL and AZ angles.

$$\mathrm{EL} = \tan^{-1}(\tan\theta \sin\phi) \tag{8.23}$$

$$\mathrm{AZ} = \tan^{-1}(\tan\theta \cos\phi) \tag{8.24}$$

Step 2: Transpose All Aperture Points to Radome Coordinates

For an AZ/EL gimbal system, the aperture point transformation to radome coordinates is found from [19; R. Matyskiela, American Electronics Laboratories, Landsdale, PA, private correspondence, 1994]:

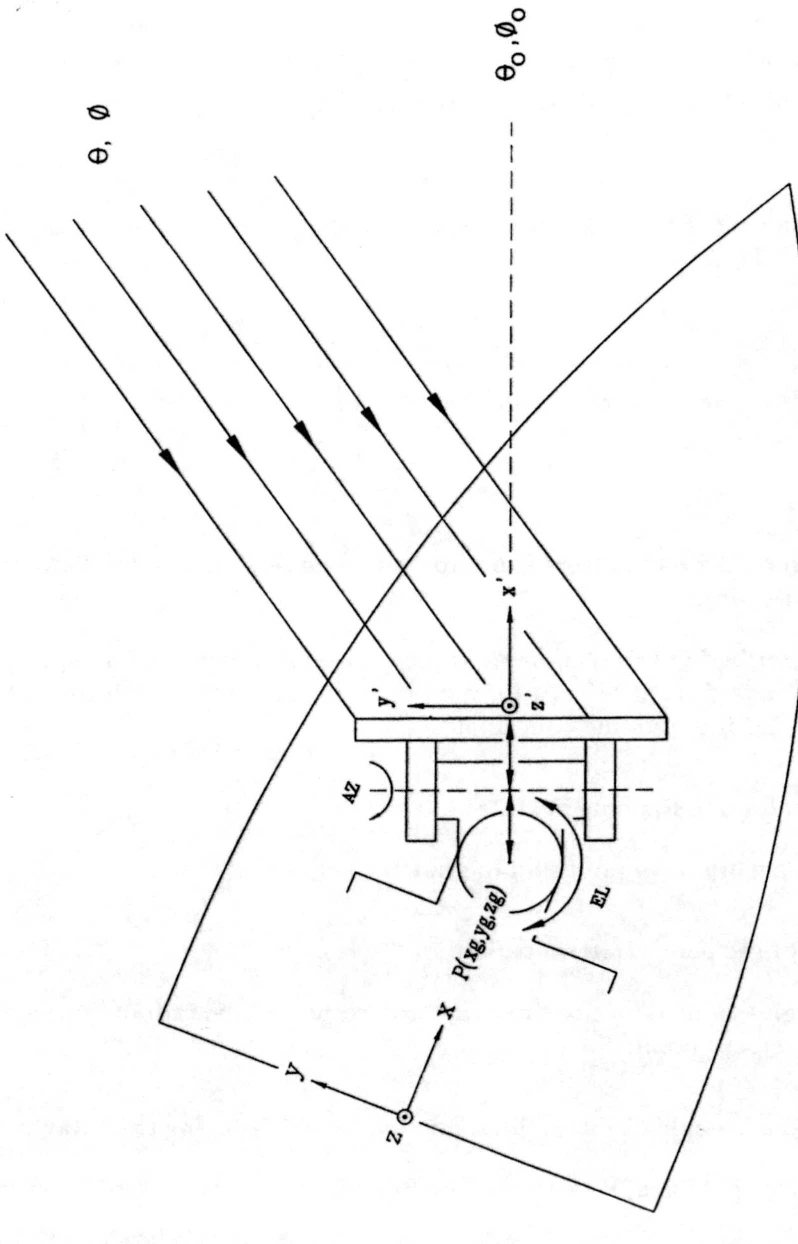

Figure 8.11 Radome and antenna coordinates used in Program RADOME.

$$\begin{bmatrix} x_i \\ y_i \\ z_i \end{bmatrix} = \begin{bmatrix} \cos AZ \cos EL & -\cos AZ \sin EL & -\sin AZ \\ \sin EL & \cos EL & 0 \\ \sin AZ \cos EL & -\sin AZ \sin EL & \cos AZ \end{bmatrix} \begin{bmatrix} \Delta_b \\ y_i' \\ z_i' \end{bmatrix} + \begin{bmatrix} \Delta_a \cos AZ \\ 0 \\ \Delta_a \sin AZ \end{bmatrix} + \begin{bmatrix} x_g \\ y_g \\ z_g \end{bmatrix} \tag{8.25}$$

Similarly, for an AZ/EL gimbal, the result, including a translation for the initial gimbal location $P(x_g, y_g, z_g)$ is

$$\begin{bmatrix} x_i \\ y_i \\ z_i \end{bmatrix} = \begin{bmatrix} \cos EL \cos AZ & -\sin EL & -\cos EL \sin AZ \\ \sin EL \cos AZ & \cos EL & \sin EL \sin AZ \\ \sin AZ & 0 & \cos AZ \end{bmatrix} \begin{bmatrix} \Delta_b \\ y_i' \\ z_i' \end{bmatrix} + \begin{bmatrix} \Delta_a \cos EL \\ \Delta_a \sin EL \\ 0 \end{bmatrix} + \begin{bmatrix} x_g \\ y_g \\ z_g \end{bmatrix} \tag{8.26}$$

Step 3: Transform All Ray Vectors from Aperture Sample Points in the Pattern Look Direction

Ray vectors described in this section are those derived from aperture sample points in the pattern look direction. Transform all these ray vectors through the antenna gimbal angles and into radome coordinates.

Step 4: Project Rays to Radome Surface

Determine all ray intercept points on the radome surface.

Step 5: Angle-of-Incidence Determination

Determine angles of incidence of these rays with respect to the radome surface normal at each intercept point.

Step 6: Compute Complex Voltage Transmission Coefficients for Each Ray

Computing each coefficient depends on the following:

- The angle of incidence of the ray relative to the surface normal at the point where it intercepts the radome wall;
- Radome wall parameters such as thickness, dielectric constant, and loss tangent of each of the layers;
- The wavelength of operation;
- The polarization of the incident wave.

The WALL program developed in Chapter 5 is utilized, but as a subroutine. Subroutine WALL makes the following assumptions: (1) The radome wall is locally plane in the vicinity of the ray intercept and (2) the radome inner and outer walls are parallel. These approximations have produced reasonably acceptable radome electrical performance predictions in the experience of the author.

Step 7: Perform an Antenna Aperture Integration

Perform an antenna aperture integration to determine the receiving port voltage from the selected pattern look direction and for each sum and difference aperture distribution. The result gives the far-field patterns, from which radome transmission loss, BSE, and BSES can be derived. The array pattern normalized to the no-radome case is given in the following formuia:

$$E_T = \frac{\sum_{m=1}^{M} \sum_{n=1}^{N} F_{mn}^a T_{mn} e^{-j(x_{mn}k_x + y_{mn}k_y + z_{mn}k_z)}}{\sum_{m=1}^{M} \sum_{n=1}^{N} F_{mn}^a} \tag{8.27}$$

where T_{mn} is the complex transmission coefficient of the radome wall for the ray associated with the mn'th aperture point, and k_x, k_y, and k_z are the propagation constants in the x, y, and z planes, respectively.

Figure 8.12 shows a computer software flow diagram for a program called RADOME, which was designed to compute the effects of a radome on a monopulse antenna. A complete listing for a BASIC program based on this flow diagram appears as Appendix D at the end of this book.

As configured, program RADOME treats tangent ogive radome shapes. You can fit this program to your particular situation by making the following choices:

1. Choose either AZ/EL or EL/AZ antenna gimbal configurations.
2. Model radome wall geometry up to 10 layers.
3. Specify the thickness, dielectric constant, and loss tangent of each of these layers.

.4 ADDITIONAL RADOME MODELING CONSIDERATIONS

his subsection discusses miscellaneous factors that may be necessary to model in a ven situation. The discussion that follows applies principally to radomes used on rborne vehicles such as aircraft or missiles.

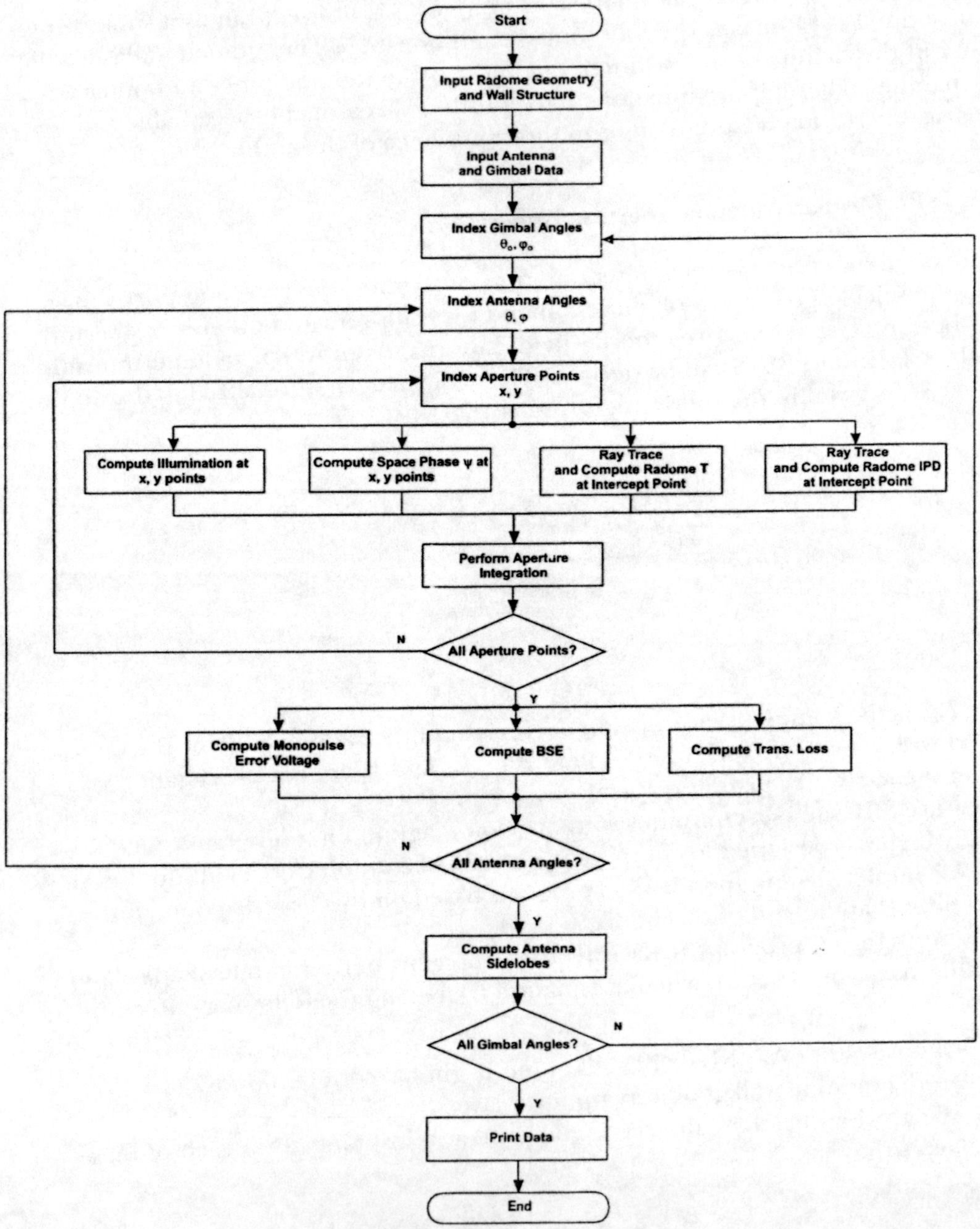

Figure 8.12 Computer software flow diagram for Program RADOME.

Rain Erosion of Radome

For high-speed flight through rain, significant radome material may erode from the outer wall, particularly in ceramic radome materials. To account for the wall thickness reduction, you must modify the radome wall thickness during the modeling process. A relationship of the material eroded per unit area (T_k) for fixed exposure time and the rain impingement angle γ_r (relative to the radome surface) can be defined for many materials:

$$T_k = K_m (V \sin\gamma_r)^{\alpha_m - 1} \tag{8.28}$$

where V is the velocity in meters per second, and K_m and α_m are the constants for a particular material. For example, slip cast fused silica (SCFS), a popular monolithic radome material, offers excellent thermal shock resistance and electrical properties. However, it has poor rain erosion resistance above Mach 2 [20]. Balageas and Hivert have published an empirical formula to calculate the material erosion rate of SCFS based on simulation test results [21]:

$$\frac{dT_k}{dt} = 10^{-9}(3.67 - 6.78D_r + 3.94D_r^2)C_r V^{6.3}(\sin\gamma_r)^{5.3} \tag{8.29}$$

where

dT_k/dt = Rate of material erosion (μm/s);
D_r = Diameter of rain drop (mm);
C_r = Rain concentration (g/cm^3).

The preceding equation calculates total erosion incurred by the radome by inputting velocity as a function of time on a flight path, and then integrating to find the total material eroded.

Equations similar to the preceding model require a knowledge of rain data. A computerized environmental analysis technique has been developed to permit the rapid evaluation of path-cumulative rainfall statistics. The analysis postulates a generic thunderstorm of circular cross-section and varying vertical intensity. At a selected altitude, the path-cumulative rainfall is evaluated along a path of arbitrary origin and direction. It is believed that this approach could provide a much needed means for quantitatively evaluating the rain environment limitations of tactical missile radomes [22].

For candidate radome materials, erosion data can be obtained by testing small samples on a high-speed sled through a man-made (simulated) rain field. Tests reported in the technical literature have been at velocities up to approximately Mach 5 (over 1.6 km per second) [23]. Silicon nitride has been sled-tested in simulated rain to determine the influence of porosity and density on the response of this material to the rain droplet impact [24]. Variations of Duroid (a glass micro fiber loaded Teflon) material have been run in sled tests [25] to determine the optimum fiber type, content, and orientation for minimum aerothermal ablation and rain erosion.

Data published in the technical literature indicate that in the special case of normal (or near-normal) impact, even with subsonic velocities, it is necessary to put a rain erosion cap at the tip of the radome. This cap is usually metallic. To combine very high rain erosion resistance with good electrical properties, a dense silicon nitride cap obtained by hot pressed sintering has been employed, as described in [26]. In radome analysis modeling, an approximation of the effects of a metallic nose cap can be obtained by nulling all rays that intercept the nose cap (geometric optics approach).

Aerothermal Heating

For supersonic homing missiles whose mission includes attack of supersonic maneuvering targets at high altitude, the guidance system absolutely requires high accuracy. As a result, the guidance engineer places a stringent requirement on the allowable BSEs caused by the radome. For this reason, many radomes are prescription ground prior to installation on the missile airframe [27].

The effects of aerodynamic heating are (1) a decrease in dielectric permittivity and (2) an increase in loss tangent. As a result, radome BSE, BSES, and losses increase with increasing radome wall temperature.

To model aerodynamic heating of the radome wall, it is often convenient to express the temperatures over the entire surface of the radome as a function of the longitudinal position and as a function of the circumferential position.

Computer programs exist that determine temperatures of an aircraft or missile radome due to aerodynamic heating. This heating is a function of trajectory, altitude profile, velocity, and maneuvers. One such program, URLIM [28], calculates temperatures for various points along a trajectory. It also computes the bending stresses at the base of the radome due to aerodynamic pressure and maneuver loads.

Thermal analysis theory is discussed in a report written by the Georgia Institute of Technology [28]. The approach generates temperature profiles through the radome material versus trajectory time. The model predicts the effects of temperature changes on electromagnetic windows for single- and multiple-layer windows. Temperature effects on expansion, complex permittivity, and complex permeability are included in the model. Performance factors include transmission, reflection, insertion phase delay, and absorption as functions of the angle-of-incidence, polarization, and temperature for a flat-plate single-or multiple-layer slab.

The philosophy followed by electromagnetic (EM) radome designers is to provide a radome having low BSEs at ambient temperature, but not at high temperatures [29]. Because radome-induced errors increase with temperature, if the magnitude of the boresight error is large enough, an oscillatory guidance system instability can result [30]. Compensation (filtering) within the guidance loop is possible to minimize the effects of the boresight error. However, compensation decreases missile responsiveness and missile effectiveness. Missile instability may not necessarily result in significant performance degradation if the errors are confined to small regions of the radome [31].

Computation of Radome Effects on Conscan Antennas

Conical scan (conscan) tracking systems are generally less complex and less expensive than monopulse systems. Using conscan, a single antenna beam is offset from the antenna axis and rotated about the antenna. For instance, consider physically rotating an offset antenna reflector or antenna feed. The modulations of a received signal within the antenna beamwidth are correlated with resolvers on the scanning mechanism to determine the position of the target relative to the antenna axis. Conscan is popular for use in small missiles.

Although conscan systems have simplicity and cost advantages relative to monopulse systems, they also have performance deficiencies. For instance, because of the offset and beam nutation, the actual beamwidth of the antenna is up to 40% greater than the minimum beamwidth possible. In some cases, the crossover of the beam is at the half-power point of the two-way antenna pattern. On antenna boresight, a net loss then results of about 3 dB of receiving signal gain [10].

In comparing data for conscan antennas with monopulse antennas of the same diameter and gimbal location within the same radome, the differences in computed BSE and radome transmission loss are very small. Therefore, a conscan antenna can be approximately modeled by a monopulse antenna of the same diameter, using the techniques of this chapter.

References

[1] Dowsett, P. H., "Cross Polarization in Radomes: A Program for Its Computation," *IEEE Trans. Aerospace Electronics Systems,* Vol. AES-9, No. 3, May 1973, pp. 421–433.

[2] Huddleston, G. K., "Radomes," Chap. 42 in *Antenna Engineering Handbook,* 3rd ed., R. C. Johnson, Ed., New York: McGraw Hill Book Company, 1992.

[3] Gulick, J. F., "Overview of Missile Guidance," *IEEE Eastcon Record,* September 1978, pp. 194–198.

[4] Sparks, R. A., "Systems Applications of Mechanically Scanned Slotted Array Antennas," *Microwave Journal,* Vol. 31, No. 6, June 1988, pp. 26–48.

[5] Lalezari, F., T. C. Boone, and J. M. Rogers, "Planar Millimeter Wave Arrays," *Microwave Journal,* Vol. 34, No. 4, April 1991, pp. 85–92.

[6] Kelly, W., "Homing Missile Guidance—A Survey of Classical and Modern Techniques," *IEEE Southcon Technical Proc.,* January 1981.

[7] Savage, P. G., "A Strapdown Phased Array Radar Tracker Loop Concept for a Radar Homing Missile," *AIAA Guidance Control Flight Conference,* August 1969, pp. 1–8.

[8] Schuchardt, J. M., and D. J. Kozakoff, "Seeker Antennas," Chap. 38 in *Antenna Engineering Handbook,* 3rd ed., R. C. Johnson, Ed., New York: McGraw Hill Book Company, 1992.

[9] Hirsch, H. L., and D. C. Grove, *Practical Simulation of Radar Antennas and Radomes,* Norwood, MA: Artech House, 1987.

[10] Skolnik, M. I., *Introduction to Radar Systems,* 2nd ed., New York: McGraw Hill Book Company, 1970.

[11] Huddleston, G. K., "Near-Field Effects on Radome Boresight Errors," *Proc. 17th Symp. on Electromagnetic Windows,* Georgia Institute of Technology, Atlanta, GA, July 1984, pp. 41–55.

[12] Hacker, P. S., and H. E. Schrank, "Range Distance Requirements for Measuring Low and Ultra Low Sidelobe Antenna Patterns," *IEEE Trans. Antennas Propagation,* Vol. AP-30, No. 5, September 1982, pp. 956–965.

[13] Hansen, R. C., "Measurement Distance Effects on Low Sidelobe Patterns," *IEEE Trans. Antennas Propagation,* Vol. AP-32, No. 6, June 1984, pp. 591–594.

[14] Mailloux, R. J., *Phased Array Antenna Handbook,* Norwood, MA: Artech House, 1994.

[15] Sokolnikoff, I. S., and R. M. Redheffer, *Mathematics of Physics and Modern Engineering,* New York: McGraw Hill Book Company, 1958.

[16] Haward, R. A., E. L. Rope, and G. Tricoles, "Radome Boresight Error and Wavefront Deformation," *Proc. 1977 IEEE Antenna and Propagation Society Symp.,* IEEE Catalog Number 77CH1217-9AP, 1977.

[17] Siwiak, K., T. Dowling, and L. R. Lewis, "Numerical Aspects of Radome Boresight Error Analysis," *Proc. 14th Symp. on Electromagnetic Windows,* Georgia Institute of Technology, Atlanta, GA, 1978.

[18] Schuchardt, J. M., D. J Kozakoff, D. O. Gallentine, T. N. Long, M. M. Hallum, and B. F. Wilson, "Automated Radome Performance Evaluation in the Radio Frequency Simulation System (RFSS) Facility at MICOM," *Proc. 15th Symp. on Electromagnetic Windows,* Georgia Institute of Technology, Atlanta, GA, June 1980.

[19] **[[AU to provide ref. (old [36])]]**

[20] Barta, J., "Rain and Sand Resistance of SCFS Radomes," *Proc. 18th Symp. on Electromagnetic Windows,* Georgia Institute of Technology, Atlanta, GA, September 1986, pp. 131–138.

[21] Balageas, D., and A. Hivert, "Rain Erosion, a Serious Problem for Slip Cast Fused Silica," *Proc. 13th Symp. on Electromagnetic Windows,* Georgia Institute of Technology, Atlanta, GA, September 1976, pp. 45–49.

[22] Crowe, B. J., "Radome Rain Damage—An Environmental Analysis Technique," *Proc. 15th Symp. on Electromagnetic Windows,* Georgia Institute of Technology, Atlanta, GA, June 1980.

[23] Frazer, R. K., "Rain Erosion Tests of Full Size Slip Cast Fused Silica Radomes at M3. 5 and M4. 8," *Proc. 17th Symp. on Electromagnetic Windows,* Georgia Institute of Technology, Atlanta, GA, July 1984.

[24] Schmitt, G. F., "Influence of Porosity and Density on the Supersonic Rain Erosion Behavior of Silicon Nitride Radome Materials," *Proc. 13th Symp. on Electromagnetic Windows,* Georgia Institute of Technology, Atlanta, GA, September 1976, pp. 37–44.

[25] Bomar, S. H., H. L. Bassett, P. R. Theobald, and J. M. Newton, "Radome Materials Evaluation," Technical Report AFAL-TR-73-222, Wright Patterson Air Force Base, Dayton, prepared by Georgia Institute of Technology, Atlanta, GA, 1973.

[26] Letson, K. N., W. G. Burleson, and R. A. Reynolds, "Influence of Angle of Incidence on the Rain Erosion Behavior of Duroid Radome Materials," *Proc. 16th Symp. on Electromagnetic Windows,* Georgia Institute of Technology, Atlanta, GA, June 1982, pp. 181–186.

[27] Weckesser, L. B., R. K. Frazer, D. J. Yost, B. E. Kuehne, G. P. Tricoles, R. Hayward, and E. L. Rope, "Aerodynamic Heating Effects on Radome Boresight Errors," *Proc. 14th Symp. on Electromagnetic Windows,* Georgia Institute of Technology, Atlanta, GA, June 1978.

[28] Frazer, R. K., "Use of the URLIM Computer Program for Radome Analysis," *Proc. 14th Symp. on Electromagnetic Windows,* Georgia Institute of Technology, Atlanta, GA, June 1978, pp. 65–70.

[29] Kuehne, B. E., and D. J. Yost, "When Are Boresight Error Slopes Excessive?," *Proc. 14th Symp. on Electromagnetic Windows,* Georgia Institute of Technology, Atlanta, GA, June 1978.

[30] Jefferson, F. L., "Minimizing Effects of Temperature Changes on Electromagnetic Windows," *Proc. 20th Symp. on Electromagnetic Windows,* Georgia Institute of Technology, Atlanta, GA, September 1992, pp. 50–55.

[31] Weckesser, L. B., "Radome Aerodynamic Heating Effects on Boresight Error," *Proc. 15th Symp. on Electromagnetic Windows,* Georgia Institute of Technology, Atlanta, GA, June 1980, pp. 97–101.

Selected Bibliography

Garnell, P., and D. J. East, *Guided Weapons Control Systems,* London: Pergamon Press, 1977.

Hayward, R. A., E. L. Rope, and G. Tricoles, "Accuracy of Two Methods for Numerical Analysis of Radome Electromagnetic Effects," *Proc. 14th Symp. on Electromagnetic Windows,* Georgia Institute of Technology, Atlanta, GA, June 1978, pp. 53– 57.

Hayward, R. A., E. L. Rope, and G. Tricoles, "Radome Boresight Error and Its Relation to Wavefront Distortion," *Proc. 13th Symp. on Electromagnetic Windows,* Georgia Institute of Technology, Atlanta, GA, September 1976, pp. 87–92.

Letson, K. N, et. al., "Rain Erosion and Aerothermal Sled Test Results on Radome Materials," *Proc. 14th Symp. on Electromagnetics Windows,* Georgia Institute of Technology, Atlanta, GA, June 1978, pp. 109–116.

Pendergrass, T. S., "Radome Analysis," Technical Report AD-A136805, Huntsville, AL: Teledyne Brown Engineering, 1983.

Appendix 8A: Program PATTERN Software Listing

——— Computes Radiation Patterns of a Monopulse Antenna ———

```
        CLS
        PI = 3.14159265#
'
'       Input Variables:
'       F                  =       Frequency (GHz)
'       WAVE               =       Wavelength (inches)
'       diameter           =       antenna diameter (inches)
'       radius             =       antenna radius (inches)
'       PHI,THETA          =       spherical coordinate angles (radians)
'       Output Variables:
'       RESUM,IMSUM        =       Real and Imaginary parts of monopulse
'                                  sum pattern (volts)
'       REDELEL,IMDEL      =       Real and imaginary parts of monopulse
'                                  elevation difference pattern (volts)
'       REDELAZ,IMDELEZ =          Real and imaginary parts of monopulse
'                                  azimuth difference pattern (volts)
'       VEL,VAZ            =       Monopulse elevation and azimuth error
'                                  voltages, respectively (volts)
' - - - - - - - - - - - - - - - - - - - - - - - - - - - - - - - - - - - -
'
'       Input data for calculations:
        F = 10
        diameter = 15
```

```
        WAVE = 11.81 / F
        radius = diameter / 2
'       Compute normalization constant:
        FIRST = 0: NORM = 0
        GOSUB 300
        FIRST = 1
'       INDEX ANTENNA ANGLES:
'       Compute principal plane antenna radiation patterns
'        Azimuth plane cut:
        DPHI = 0
        LPRINT "Azimuth Principal Plane Pattern"
        LPRINT ""
        LPRINT "THETA   PHI   SUM   VAZ   VEL"
        LPRINT "": LPRINT ""
'
        FOR DTHETA = 0 TO 150 STEP 5
        PHI = DPHI * PI / 180
        THETA = (DTHETA / 10) * PI / 180
        GOSUB 300
        SUM = SQR(RESUM ^ 2 + IMSUM ^ 2) / NORM
        SUM = 20 * LOG(SUM) / LOG(10)
'       Compute Monopulse Error Voltages
        VAZ = IMDELAZ / RESUM
        VEL = IMDELEL / RESUM
        LPRINT USING " +###.#  +###.#   +###.#   +###.#   +###.#";
DTHETA; DPHI; SUM; VAZ; VEL
        NEXT DTHETA
        LPRINT "": LPRINT ""
'       Elevation plane cut
        DPHI = 90
        LPRINT "Elevation Principal Plane Pattern"
        LPRINT ""
        LPRINT "THETA   PHI   SUM   VAZ   VEL"
        LPRINT ""

        FOR DTHETA = 0 TO 150 STEP 5
        PHI = DPHI * PI / 180
        THETA = (DTHETA / 10) * PI / 180

        GOSUB 300
        SUM = SQR(RESUM ^ 2 + IMSUM ^ 2) / NORM
        SUM = 20 * LOG(SUM) / LOG(10)
'       Compute Monopulse Error Voltages
        VAZ = IMDELAZ / RESUM
        VEL = IMDELEL / RESUM
        LPRINT USING " +###.# +###.# +##.## -=#.## +##.##"; DTHETA; DPHI;
SUM; VAZ; VEL
        NEXT DTHETA

        LPRINT CHR$(12)
        END
```

```
300 '   Aperture Integration Subroutine
        RESUM = 0: IMSUM = 0: REDELAZ = 0: IMDELAZ = 0: REDELEL = 0:
 IMDELEL = 0
        FOR I = 0 TO 20
                x = ((I - 10) / 10) * radius
        FOR J = 0 TO 20
                y = ((J - 10) / 10) * radius

        IF x = 0 AND y = 0 THEN
                FSUM = 0: FDELAZ = 0: FDELEL = 0
                GOTO 100
        END IF
                r = SQR(x ^ 2 + y ^ 2)
                IF r > radius THEN GOTO 200

'       Compute Illumination at x,y points:
                FSUM = COS(.5 * PI * r / radius)
                FDELAZ = SIN(PI * x / radius)
                FDELEL = SIN(PI * y / radius)
'       Element Array Factor:
                FEL = COS(.5 * PI * x / radius)

100 '   CONTINUE

        IF FIRST = 0 THEN
                NORM = NORM + FSUM * FEL
                GOTO 200
        END IF

'       Compute Space Phase at x,y points:

        PSI = (x * SIN(THETA) * COS(PHI) + y * SIN(THETA) * SIN(PHI)) * 2
 * PI / WAVE

'       Perform Aperture Integration:

        RESUM = FSUM * FEL * COS(PSI) + RESUM
        IMSUM = FSUM * FEL * SIN(PSI) + IMSUM

        REDELAZ = FDELAZ * FEL * COS(PSI) + REDELAZ
        IMDELAZ = FDELAZ * FEL * SIN(PSI) + IMDELAZ

        REDELEL = FDELEL * FEL * COS(PSI) + REDELEL
        IMDELEL = FDELEL * FEL * SIN(PSI) + IMDELEL

200 '   CONTINUE

        NEXT J
        NEXT I

        RETURN
```

Chapter 9

Other Radome-Enclosed Antenna Types

Increasingly, radome-enclosed antennas are finding new applications in technology areas that include, but are not limited to, satellite navigation and communications, telecommunications, and avionics. This chapter discusses radome analysis methods for spiral antennas, parabolas and other telecommunications antennas, air-supported radomes, metal space frame radomes, and radome-enclosed phased-array antennas.

9.1 SPIRAL ANTENNAS

A planar spiral antenna as illustrated in Figure 9.1 is a very broadband, circularly polarized antenna element. Planar spirals are popular for use in antiradiation homing (ARH) missile guidance systems [1] or in electronics countermeasure (ECM) direction-finding systems on aircraft. A complete discussion of this category of antennas appears in Chapter 3.

Spiral antennas typically use printed circuit technology. However, this limits their use to low-power transmitting or receiving applications. Thus, spirals are usually used for passive applications. Current tolerances obtained with this type of fabrication technology permit spiral electrical operation to at least the K_a-band on the high end of the spectrum. Spiral antennas are available to at least L-band on the low end of the spectrum. Furthermore, certain loading techniques can reduce size or permit operation at even lower frequencies [2].

The typical antenna half-power beamwidth (HPBW) is up to 90 degrees. This wide, instantaneous field of view makes these antennas attractive for many body fixed missile guidance applications [3]. Spiral antennas can by used in arrays of two or more elements in small interferometers [4]. Missile guidance systems using these are usually fixed-mounted (nongimballed) with pursuit navigation algorithms to

Figure 9.1 A broadband, cavity-backed, planar Archimedean spiral antenna. (Photo courtesy of Tracor Aerospace Electronic Systems.)

obtain line-of-sight data to the target [5]. Figure 9.2 illustrates several array geometries for reference.

For the no-radome condition, the principal plane sum and difference patterns for the two-element, spiral antenna array are given by the following formulas:

$$\Sigma = \cos\left(\frac{\pi d_a}{\lambda_o}\sin\theta\right)\cos(\theta) \tag{9.1}$$

$$\Delta = \sin\left(\frac{\pi d_a}{\lambda_o}\sin\theta\right)\cos(\theta) \tag{9.2}$$

where

θ = Angle measured from the antenna boresight;
d_a = Center-to-center antenna element spacing;
λ_o = The operating wavelength in free space.

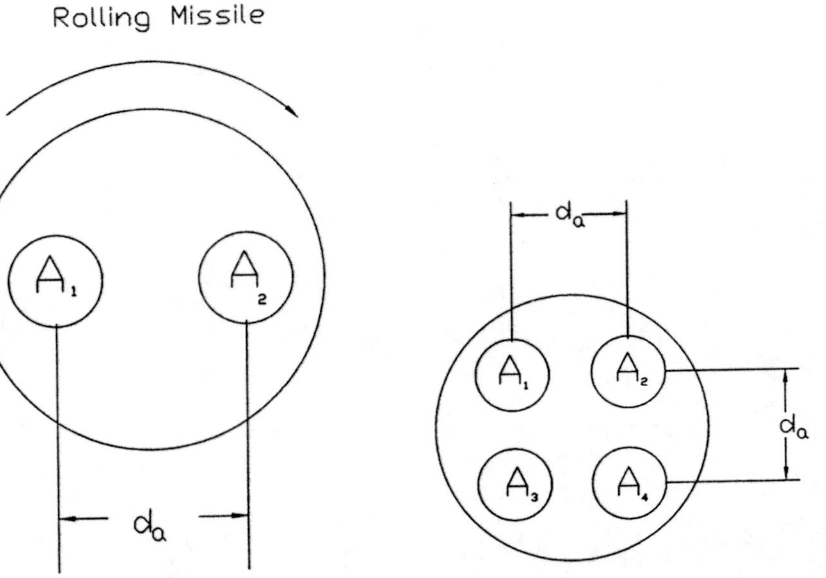

Figure 9.2 Spiral antenna array geometries: (a) two-element (rolling missile) and (b) four-element square.

Figure 9.3 shows a microwave circuit to resolve the received antenna signals from a four-element spiral array into monopulse sum and difference patterns.

Placing a radome over a spiral antenna array results in an apparent boresight error (BSE) because of the wavefront phase modification over the antenna apertures. These effects need to be carefully considered during the selection of a radome design since they will influence the seeker system accuracy performance [6–8]. Broadband ARH or ECM antennas generally require multilayer radome walls to obtain a required radome bandwidth commensurate with desired antenna performance. Half-wave wall radomes generally have a useful bandwidth of only 5% to 20%, depending on the dielectric constant and the radome fineness ratio. Higher order radome walls that are integral multiples of half-wave thick walls have a considerably narrower bandwidth compared to the half-wave wall. They are limited only to very narrowband antenna applications.

When using a spiral antenna array, the relationship between BSE [and boresight error slopes (BSESs)] and the miss distance of a missile is not straightforward. Sometimes we quantify these issues only by hardware-in-the-loop (HWIL) simulation testing that takes into account a variety of scenarios and flight conditions [9,10]. Another method is based on open-loop testing in a suitable anechoic chamber facility. This method requires a precalibration of the monopulse difference channels' sensitivity in volts per degree without the radome in place. Subsequently, the difference channel voltage characteristics are measured with the

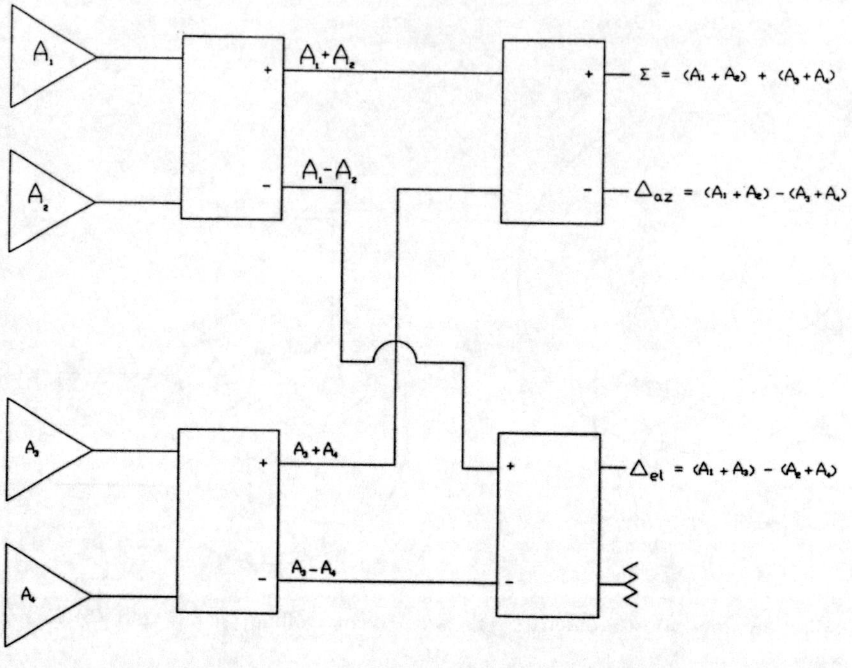

Figure 9.3 Microwave circuit to resolve the received antenna signals from a four-element spiral array into monopulse sum and difference patterns.

radome placed over the antenna array. The following figures are typical for many radome types [3]:

- A BSE requirement of equal to or less than 10 mrad;
- A BSES requirement of equal to or less than 0.05 deg/deg;
- A maximum transmission loss of −1 dB.

The following method is useful for estimating the BSE characteristics of the spiral antenna array. From Figure 9.3, rewrite the monopulse sum and difference formulas to incorporate a radome transmission coefficient associated with each antenna, based on the simple ray trace depicted in Figure 9.4.

$$\Sigma = (T_{w_1} A_1 + T_{w_2} A_2) + (T_{w_3} A_3 + T_{w_4} A_4) \tag{9.3}$$

$$\Delta_{az} = (T_{w_1} A_1 + T_{w_2} A_2) - (T_{w_3} A_3 + T_{w_4} A_4) \tag{9.4}$$

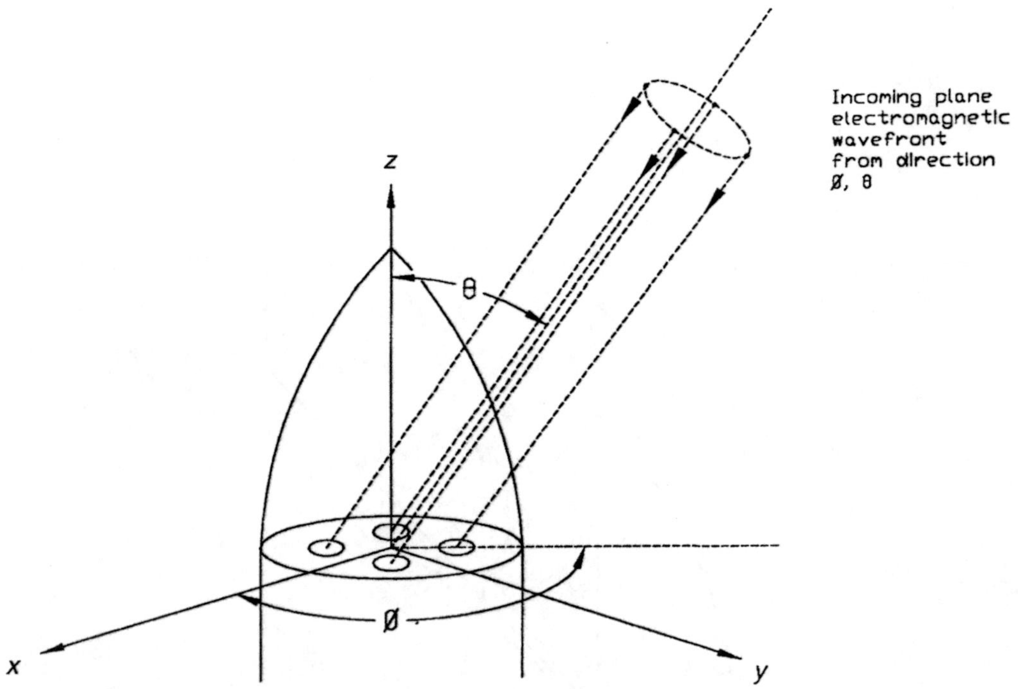

Figure 9.4 Simple array ray trace technique.

$$\Delta_{el} = (T_{w_1} A_1 + T_{w_3} A_3) - (T_{w_2} A_2 + T_{w_4} A_4) \tag{9.5}$$

In these equations, the signal amplitudes of each of the spiral antennas $|A_i|$ are normally all about equal. However, the phase associated with each varies in the array. The transmission coefficients, T_{w_i}, can be evaluated by the wall construction, material properties, polarization, wavelength, and the angle-of-incidence of a ray with respect to the surface normal at the intercept point.

A more accurate method of obtaining estimates of radome impact on the performance of the spiral array is based on equivalent transmission coefficients. Assuming a receiving mode, the surface integration shown in Figure 9.5 illustrates the method. The procedure involves the physical optics method discussed in Section 6.3, which models an incoming plane wavefront as an ensemble of Huygens's sources. The external reference plane is used to transform points from the ensemble of Huygens's sources to a point centered on each spiral antenna aperture [11]. In other words, we trace rays coming from points on the external reference plane, incorporating the radome transmission coefficient to modify each ray. From this process, we derive an equivalent transmission coefficient in a given look direction θ, ϕ:

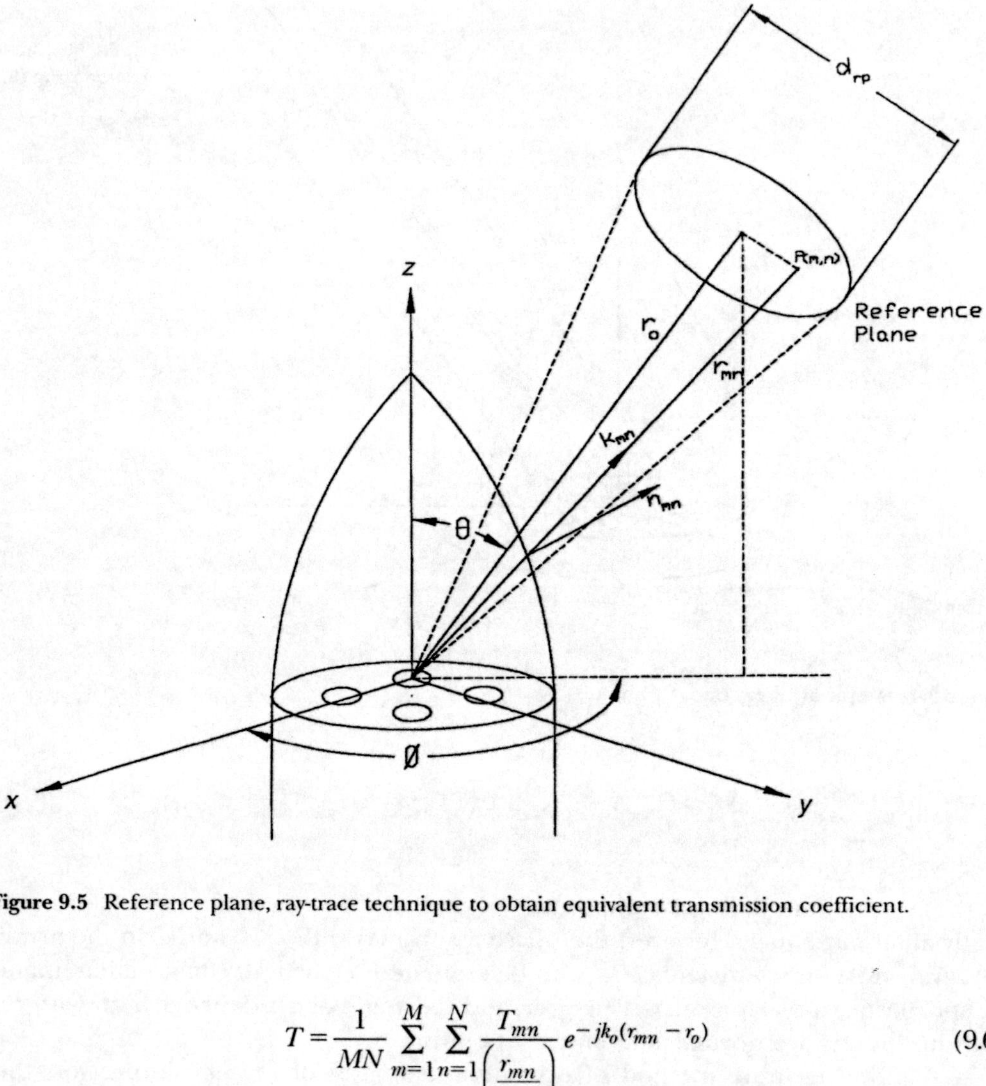

Figure 9.5 Reference plane, ray-trace technique to obtain equivalent transmission coefficient.

$$T = \frac{1}{MN} \sum_{m=1}^{M} \sum_{n=1}^{N} \frac{T_{mn}}{\left(\frac{r_{mn}}{r_o}\right)} e^{-jk_o(r_{mn} - r_o)} \tag{9.6}$$

where the m and n indices correspond to points on the external reference plane.

Multimode Spiral Antennas

If the diameter of a spiral antenna is sufficiently large, higher order modes may exist. These correspond to active ring circumferences corresponding to $P\lambda$ for the P'th mode, where P is an integer. The electrical phase of the currents on these active rings varies from 0 to $2\pi P$ as we progress around the radiating current band. All modes $P > 1$ have a radiation pattern null on boresight. Corzine and Mosko [12] present an in-depth treatment of how the first two modes might be used to generate monopulse sum and difference signals from a single antenna. A circuit for processing the first two modes of a four-arm spiral into monopulse sum and difference signals was presented and discussed in Chapter 3.

Very little technical literature exists on the radiation patterns of cavity-backed, planar spiral antennas. However, Reedy et al. [13] have recently developed a combined method of moments (MOM) and general theory of diffraction (GTD) technique.

A simplified method that is proposed in the following involves performing a numerical integration of the ring current band corresponding to the spiral radiation mode of interest, as illustrated in Figure 9.6. This method will yield a reasonable approximation to the antenna radiation pattern in the presence of a radome.

For the P'th current ring, if we assume that there are a total of M sample points around the ring, the differential phase shift between current sample points is

$$\phi_M = \frac{2\pi P}{M} \tag{9.7}$$

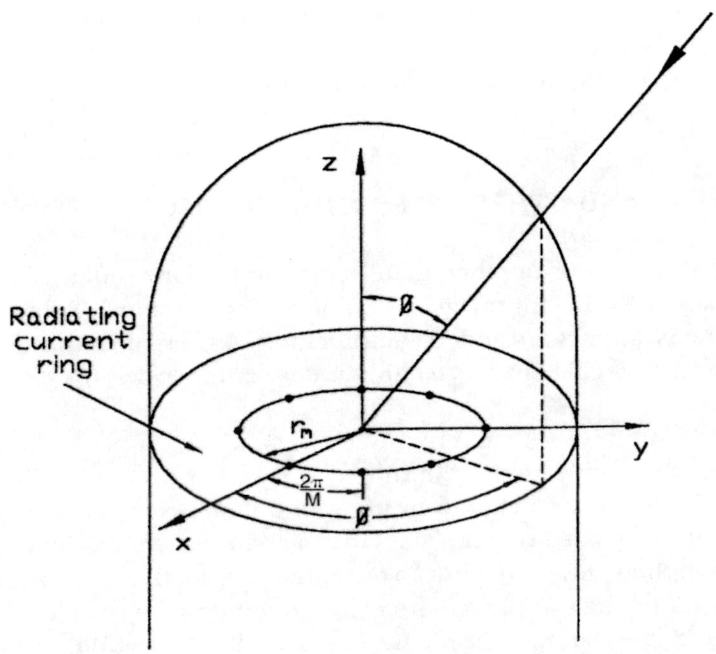

Figure 9.6 Multimode planar spiral in a hemispherical radome.

and the phase shift at sample point m is

$$\phi_m = m\phi_M = \frac{m2\pi P}{M} \tag{9.8}$$

The array factor for the P'th current ring is then broken into ϕ and θ field components:

$$E_x^{AF} = \sum_{m=0}^{M} \left(-\sin\left(\frac{m}{M}2\pi\right)e^{-j\phi_m}\right)T_m(\theta,\phi)e^{jk(x_m \sin\theta\cos\phi + y_m \sin\theta\sin\phi)} \tag{9.9}$$

$$E_y^{AF} = \sum_{m=0}^{M} \left(\cos\left(\frac{m}{M}2\pi\right)e^{-j\phi_m}\right)T_m(\theta,\phi)e^{jk(x_m \sin\theta\cos\phi + y_m \sin\theta\sin\phi)} \tag{9.10}$$

In the

$T_m(\theta, \phi)$ = Complex radome wall voltage transmission coefficient associated with the m'th sample point for a ray incident from the θ, ϕ direction;

x_m = x coordinate of m'th point on P'th current ring (m)
= $r_p \cos \phi_m$;

y_m = y coordinate of m'th point on P'th current ring (m),
= $r_p \sin \phi_m$;

r_p = Radius of P'th current ring (m),
= $P\lambda/2\pi$

9.2 PARABOLAS AND OTHER TELECOMMUNICATIONS ANTENNAS

Since World War II, the number of telecommunications antenna radomes has steadily increased. As the advantages of radomes become more widely recognized, the rate at which radomes are added continues to grow. To meet this growing need, it is important for specific analysis techniques to be developed for telecommunications radomes.

For telecommunications antennas such as parabolas, a planar fabric or membrane radome is generally used, as illustrated in Figure 9.7. These radomes are electrically thin. The losses are held low by using thin, low-loss material such as Teglar, a fiberglass material coated on both sides with Teflon that sheds water readily [14].

Because of their close proximity to the antenna feed, these planar radomes can introduce small increases in the antenna voltage standing wave ratio (VSWR) [15]. For a planar radome like the one shown in Figure 9.7, the VSWR increase can be reduced by tilting the radome a few degrees (moving the top further from an antenna aperture) to prevent specular return to the feed horn [16].

Figure 9.7 Electrically thin radome over telecommunications antenna. (Photograph courtesy of Andrew Corporation.)

9.3 AIR-SUPPORTED RADOMES

Many of the earliest ground-based radomes were hemispherical in shape and were supported by air pressure from within. A nonporous, waterproof dielectric fabric was placed over the antenna system, sealed around the base, and then inflated to clear the antenna. This fabric was electrically thin, and its loss tangent and thickness did not result in any significant transmission loss at the microwave frequencies where the system was used.

The hemispherical shape of these structures offers advantage in high winds. For instance, precipitation particles blow around a sphere rather than impinge on it. Therefore, snow, frozen precipitation, or rain is not readily deposited on the spherical surface. With these radome shapes, the incidence angles are usually less than 30 degrees. Such radomes are called *normal incidence radomes* [17]. The transmission properties of wall structures commonly used in this type of radome are relatively independent on polarization and incidence angles between 0 and 30 degrees, particularly if small tolerances on thickness variations are maintained and a narrow frequency band of operation is specified.

Early materials used for air-supported radomes were neoprene-coated nylon, hypalon-coated Dacron, and Teflon-coated fiberglass [18]. Hemispherical air-supported radomes greater that 70m in diameter are in service today [19].

These are the advantages of air-supported radomes:

- Low cost;
- Good electromagnetic transparency over very wide frequency bandwidths;
- Relative ease of erection and dismantling.

These characteristics make air-supported radomes particularly suitable for mobile stations. However, there are a few disadvantages to air-supported radomes. For instance, their life is limited by the constant flexing of the material in the wind. Furthermore, maintenance of the internal pressure in high winds can be difficult. Maintenance is costly and required at frequent intervals for the overall structure and air pressure system.

Since air-supported radomes are electrically thin, monolithic wall structures, the analysis for transmission loss can readily be done, using the WALL computer code discussed in Chapter 5. Other than transmission loss, these radomes generally have negligible effect on antenna sidelobes or depolarization.

9.4 METAL SPACE FRAME RADOMES

Rigid, self-supporting, space frame radomes overcome many limitations of air-supported radomes. As illustrated in Figure 9.8, these radomes consist of a three-di-

Figure 9.8 (a) Metal space frame radome (20.7-m diameter). (Photo courtesy of ESSCO.) (b) Impedance-matched 18.6-m diameter dielectric space frame radome. (Photo courtesy of Antennas for Communications, Inc.)

mensional lattice of primary, load-bearing, metallic members, which carry the main loads or stresses. The framework is typically steel or aluminum members. This type of design is sometimes called a *geodesic dome,* in which a spherical structure is built from simple geometrical shapes.

Geodesic dome design computer software is available in the form of DOME, a freeware DOS utility developed for generating the coordinates of a geodesic dome or sphere [20]. This zip archive includes DOS executable C++ source code. Of all the types of surface structures that have been used, the most common for large ground radomes (greater than 15m in diameter) has been the space frame radome [21].

The primary electromagnetic performance considerations to integrate into the design of a space frame radome are (1) dielectric wall transmission loss, (2) frame blockage loss, and (3) pointing errors. These considerations are discussed in the following subsections.

Figure 9.8 (continued)

Dielectric Wall Losses

The radome transmission loss in decibels is composed of two major loss contributors.

$$L = L_W + L_B \tag{9.11}$$

The first and typically the smallest contributor, L_W, is the ordinary insertion loss associated with the signal propagation through the radome wall. In this equation, it is expressed in decibels. The second loss contributor, L_B, comes from signals scattering off the frame structure. It is a blockage loss term, also expressed in decibels. For most frequency bands, the blockage loss component is larger than the insertion loss of the wall.

The dielectric radome material covering the walls of the dome may be an electrically thin, waterproof fabric, pulled tightly over the surface defined by the metal space frame. The electromagnetic insertion loss of the membrane is governed primarily by (1) the dielectric properties of the membrane and (2) the ratio of membrane thickness to operating wavelength. For the smallest possible loss, it is important to make the membrane as thin as possible.

An alternative to fabric is individual, flat, triangular panels, comprising an A-sandwich radome wall with an internal low-density foam core. The WALL program of Chapter 5 can also be applied here to compute loss component L_W through the dielectric walls. Note that with decreasing wavelength, the loss due to the walls generally increases, whereas the loss due to the radome frame remains substantially constant.

Frame Blockage Losses

In situations involving blockage by antenna feed supports or other obstacles, it has been customary to assume that the blocking satisfies geometric optics. This assumption leads to a simplified formula for blockage loss in decibels for a metal space frame radome [21]:

$$L_B = 10\log\left\{\sqrt{3}\,\frac{w_B}{L_B}\left[1+\frac{h_B}{w_B}\,\frac{\sin^{-1}\eta_A\sqrt{1-\eta_A^2}}{\eta_A^2}\right]\right\} \tag{9.12}$$

where

w_B = Beam width;
L_B = Beam length;
h_B = Beam dimension in the radial direction;
η_A = Antenna diameter-to-radome diameter ratio.

This equation can be rewritten in the following manner:

$$L_B = 10\log\left\{\sqrt{3}\,\frac{w_B}{L_B}\left[2+\frac{h_B}{w_B}\,G(\eta_A)\right]\right\} \tag{9.13}$$

In this equation, we define the optical blockage function via the next formula:

$$G(\eta_A) = \frac{\sin^{-1}\eta_A - \eta_A\sqrt{1-\eta_A^2}}{\eta_A^2} \tag{9.14}$$

This optical blockage function is plotted in Figure 9.9.

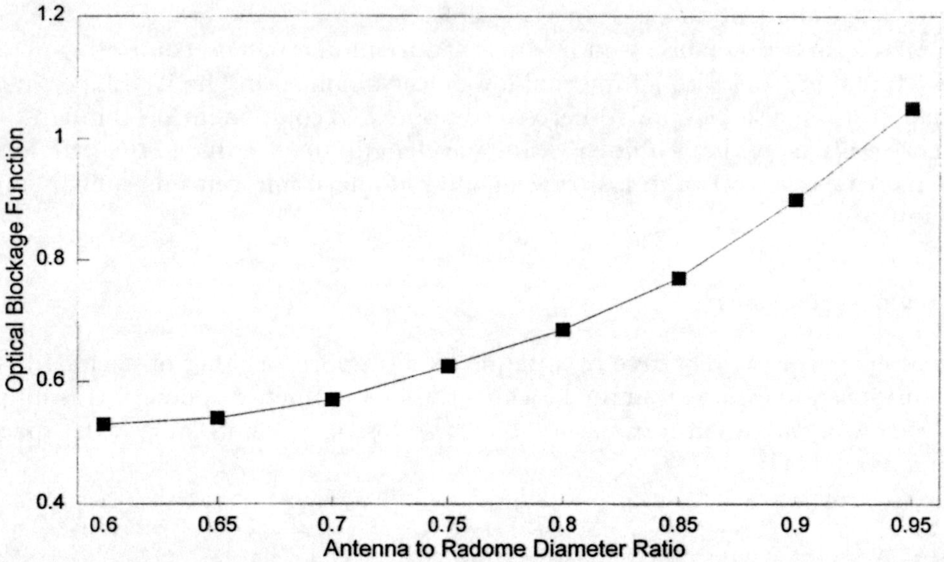

Figure 9.9 Optical blockage function versus antenna diameter-to-radome diameter ratio [21].

Pointing Errors

The pointing errors introduced by space frame radomes are extremely small, usually less than 0.1 mrad.

A uniform space frame geometry theoretically can produce no BSE. As the geometry departs from perfect uniformity, the first significant effect is BSE. For well-designed radomes, this appears to be the only significant effect. Boresight errors can result from (1) imperfect uniformity of the space frame geometry as designed and/or (2) tolerance or manufacturing variations.

To minimize BSE pattern distortion of the mainlobe and close-in sidelobes, the metallic frame geometry should have two characteristics:

1. Be as uniform as possible. In other words, the blocking area ratio should be the same in all parts of the radome. The radome icosahedral geometry shown in Figure 9.10 is one that achieves the latter requirement quite effectively [22].
2. Be as random as possible. The more random the space frame geometry, the lower the BSE.

Dielectric Space Frame Radomes

The framework of a dielectric space frame radome typically consists of plastic, as opposed to steel or aluminum members. These radomes have several advantages

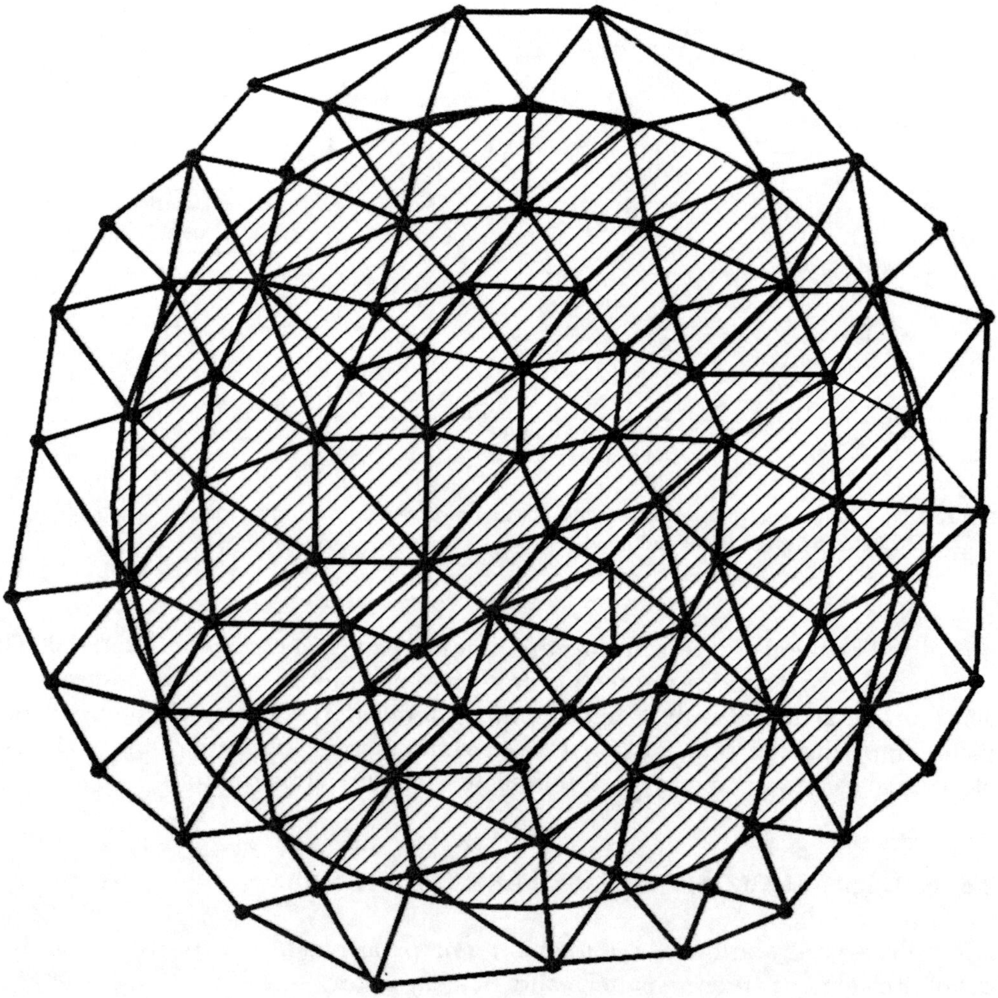

Figure 9.10 Projection of the radome onto the antenna aperture [22].

over other ground-based radome types. The radome walls may be constructed as a thin membrane, solid laminate, or as a multilayer sandwich with internal foam cores. For the thin membrane wall, radome panel flanges have dimensions capable of supporting the environmental wind loads. Often a two- or three-layer sandwich wall is used where thermal insulation or where high-frequency and low-frequency loss requirements prevail.

The dielectric beams or struts act as dielectric loads. The radome engineer can add inductive elements to impedance-match a capacitive circuit load over the required bandwidth. These inductive elements might be implemented via

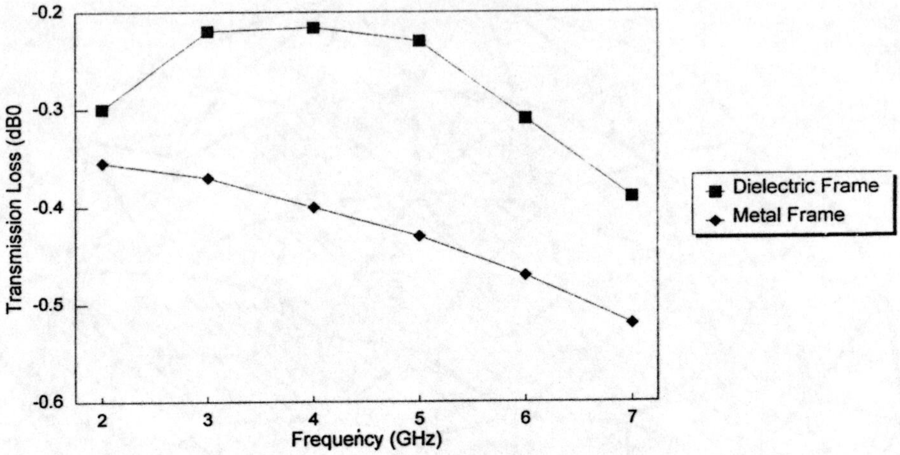

Figure 9.11 Comparison of transmission loss of a common metal space frame radome compared to a dielectric space frame radome [23].

metallized elements dispersed into the radome panel fabric or wall material. The resulting design provides reduced overall radome transmission loss, compared to what can be achieved with a metal frame radome [23]. Figure 9.11 shows typical data comparing the transmission loss of a metal space frame radome with a dielectric space frame radome of the same size and construction.

9.5 RADOME-ENCLOSED, PHASED-ARRAY ANTENNAS

For a phased-array antenna, the total antenna pattern can be expressed as a product of the element antenna pattern and an array factor:

$$E(\theta,\phi) = E_e(\theta,\phi)E^{AF} \tag{9.15}$$

The most commonly used element radiators for phased-array antennas are as follows:

- Dipoles;
- Slots in waveguide;
- Open-ended waveguides;
- Small horn antennas;
- Spiral antennas;
- Microstrip patch or disk radiator elements.

A corporate feed for a phased-array antenna is shown in Figure 9.12.

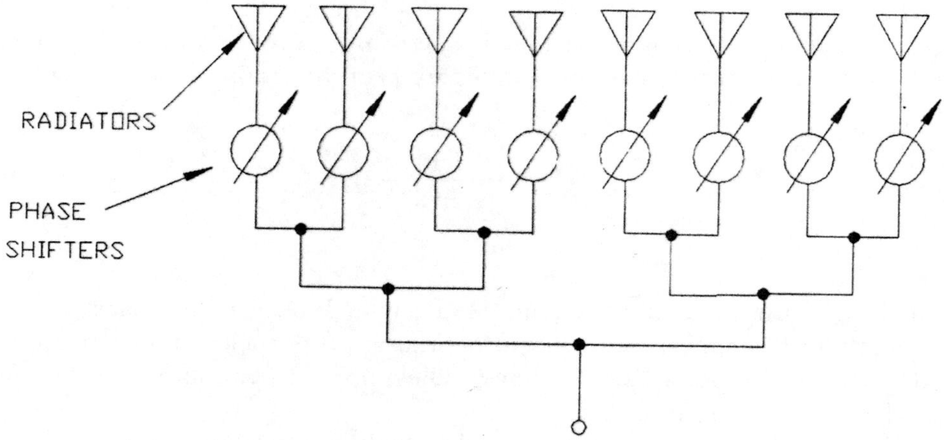

Figure 9.12 Corporate feed for a phased-array antenna.

For a square array in the *x-y* plane, assume the following:

(1) The index on elements in the *x* direction is *m*, and its greatest value is *M*; and (2) the index on elements in the *y* direction is *n*, and its greatest value is *N*.

If the element amplitude weighting distributions, B_m and C_n, are selected independently in the principal planes of the antenna aperture, a separable amplitude distribution results so that $A_{mn} = B_m C_n$. The array factor can then be written as a product of two independent variables *u* and *v* [24]:

$$E^{\mathrm{AF}} = \sum_{m=0}^{M} I_{mn} e^{-jk_o m d_x u} \sum_{n=0}^{N} e^{-jk_o n d_y v} \tag{9.16}$$

The following definitions for *u* and *v* apply in the previous equation, where the desired phased-array beam pointing angle is θ_o, ϕ_o:

$$u = \sin\theta \cos\phi \tag{9.17}$$

$$v = \sin\theta \sin\phi \tag{9.18}$$

Asymmetric aperture distributions are not used except for generating shaped antenna beams. For simplicity, only symmetric aperture distributions are assumed in this chapter. Specifically, an odd number of elements, M, along the x axis, and N elements along the y axis are assumed. The center of the array is always taken at $x = 0$, $y = 0$. Element spacings are assumed to meet the grating lobe criteria:

$$\frac{d}{\lambda_o} \leq \frac{1}{1 + \sin\theta_{max}} \tag{9.19}$$

where θ_{max} is the maximum antenna scan angle. Interelement spacings of one wavelength at the highest frequency of operation do not allow any scan. Interelement spacings of about a half wavelength allow phased-array antenna scan angles through about 80 degrees.

The interelement phase shift needed to position the beam to the pointing angle θ_o, ϕ_o, assuming a simple pencil-beam antenna pattern, is

$$u_o = \sin\theta_o \cos\phi_o \tag{9.19}$$

$$v_o = \sin\theta_o \sin\phi_o \tag{9.21}$$

Beam scanning with planar antenna arrays is accomplished by introducing a linear phase shift along both principal array coordinates, according to the preceding relationships. Two types of electronic phase shifters are currently being used in phased-array antennas: the ferrite and the diode phase shifter. The selection of the best type for a design depends on frequency and RF power handling required. A comprehensive discussion appears in Tang and Burns [25].

The current excitation at the mn'th antenna element can be expressed as follows:

$$I_{mn} = A_{mn} e^{jk(md_x \sin\theta_o \cos\phi_o + nd_y \sin\theta_o \sin\phi_o)} \tag{9.22}$$

In this form of beam steering, the required phase is the sum of a row phase and a column phase term. We can use this result to define the array factor that results with a radome in place over the phased array:

$$E^{AF} = \sum_{m=0}^{M} A_{mn} T_{mn}(\theta, \phi) e^{-jk_o md_x(\sin\theta\cos\phi - \sin\theta_o\cos\phi_o)} \sum_{n=0}^{N} e^{-jk_o nd_y(\sin\theta\sin\phi - \sin\theta_o\sin\phi_o)} \tag{9.23}$$

where $T_{mn}(\theta, \phi)$ represents the complex radome voltage transmission coefficient at the mn'th point associated with a ray incident from the θ, ϕ direction.

The array factor can be computed for any arbitrary A_{mn} distribution. For instance, Figure 9.13 shows a method of achieving simultaneous sum and difference monopulse beams for a phased-array antenna. The signals from a pair of radiating elements located symmetrically opposite from the centerline of the array are combined in a matched hybrid (magic T) to form sum and difference antenna signals. The sum signals from all the pairs across the array aperture are then combined in a power combiner network to form a sum beam.

The desired amplitude distribution for the sum beam can be achieved by proper weighting of the signals in the combiner network. The difference signals from all the pairs are combined in a separate power combiner network in order to form a difference beam. Amplitude weighting in the two combiner networks can take several forms to obtain low sidelobes for both the sum and the difference antenna beams. For instance, a Bayliss type of distribution produces low sidelobes for a difference beam [26]. For a $\cos^q$ type of sum channel distribution, for a square array the first sidelobes correspond to -23 and -32 dB for q equal to 1 and 2, respectively.

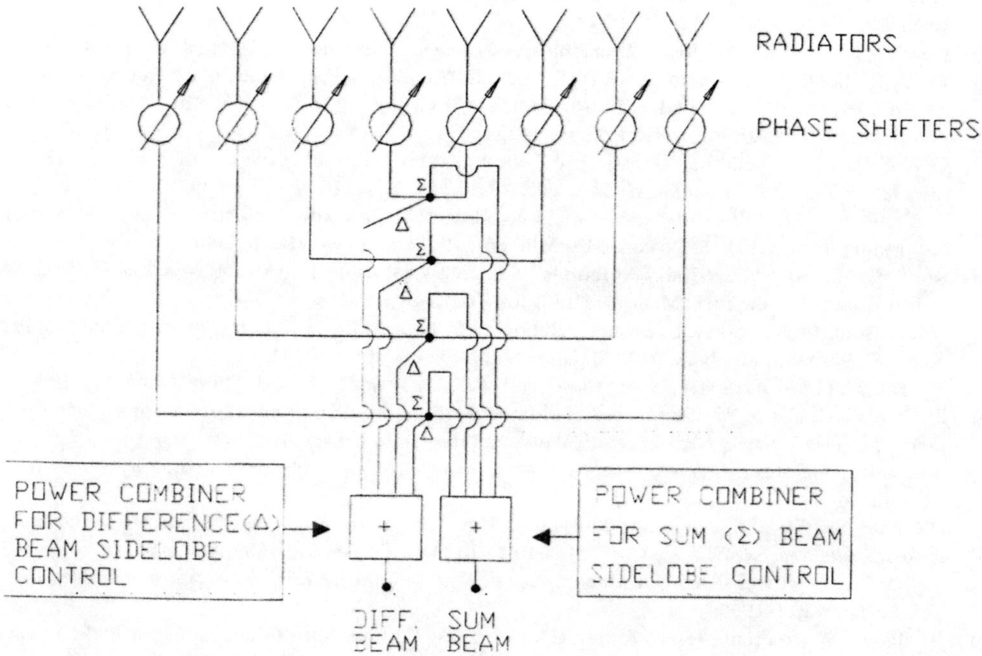

Figure 9.13 Phased-array monopulse feed with independently controlled sum and difference beams [25].

References

[1] Dyson, J. D., "Multimode Logarithmic Spiral Antennas," *National Electronics Conf.*, Vol. 17, October 1961, pp. 206–213.

[2] Morgan, T. E., "Reduced Size Spiral Antenna," *IEEE Ninth European Microwave Conf. Proc.*, UK, September 1979, pp. 181–185.

[3] Schuchardt, J. M., D. J. Kozakoff, and M. M. Hallum, "Seeker Antennas," Chap. 38 in *Antenna Engineering Handbook*, 2nd ed., R. C. Johnson and H. Jasik, Eds., New York: McGraw Hill Book Company, 1988.

[4] Schuchardt, J. M., and W. O. Purcell, "A Broadband Direction Finding Receiving System," *Martin Marietta Interdivision Antenna Symp.*, Orlando, FL, August 1967, pp. 1–14.

[5] E. R. Feagler, "The Interferometer as a Sensor for Missile Guidance," *IEEE Eastcon Record*, September 1978, pp. 203–210.

[6] Marales, G., "Simulation of Electrical Design of Streamlined Radomes," *AIAA Summer Computer Simulation Conf.*, Toronto, Canada, July 1979, pp. 353–354.

[7] Ossin, A., "Millimeter Wavelength Radomes," Report AFML-TR-79-4076, Wright Patterson Air Force Base, Dayton, OH, July 1979.

[8] Siwak, K., et al., "Boresight Errors Induced by Missile Radomes," *IEEE Trans. Antennas Propagation*, Vol. AP-27, November 1979, pp. 832–841.

[9] Russell, R. F., and S. Massey, "Radio Frequency System Simulator," *AIAA Guidance Conf.*, August 1972, pp. 72–861.

[10] Sutherlin, D. W., and C. L. Phillips, "Hardware-in-the-Loop Simulation of Antiradiation Missiles," *IEEE Southeastcon Proc.*, Clemson, SC, April 1976, pp. 43–45.

[11] Israel, M., I. Kotlarenko, J. Shapira, and S. Marcus, "A Reference Plane Method for Antenna Radome Analysis," *Proc. 15th Symp. on Electromagnetics Windows*, Georgia Institute of Technology, Atlanta, GA, 1980.

[12] Corzine, R. G., and J. A. Mosko, *Four Arm Spiral Antennas*, Norwood, MA: Artech House, 1990.

[13] Reedy, C. J., M. D. Deshpande, and D. T. Fraleeh, "Analysis of Elliptically Polarized Cavity Backed Antennas Using a Combined FEM/MOM/GTD Technique," NASA CR-198197, National Aeronautics and Space Administration, August 1995.

[14] C. A. Sellar, Jr., "Preliminary Testing of Teflon as a Hydrophobic Coating for Microwave Radomes," *IEEE Trans. Antennas Propagation*, Vol. AP-27, No. 4, July 1979, pp. 555–557.

[15] Redheffer, R. M., "The Interaction of Microwave Antennas with Dielectric Sheets," Radiation Laboratory Report 484-18, Boston, MA: MIT Lincoln Laboratory, March 1946.

[16] Knop, C. M., "Microwave-Relay Antennas," Chap. 30 in *Antenna Engineering Handbook*, 3rd ed., R. C. Johnson, Ed., New York: McGraw Hill Book Company, 1993.

[17] Huddleston, G. K., and H. L. Bassett, "Radomes," Chap. 44 in *Antenna Engineering Handbook*, 3rd ed., R. C. Johnson, Ed., New York: McGraw Hill Book Company, 1994.

[18] Skolnik, M. I., *Introduction to Radar Systems*, 2nd ed., New York: McGraw Hill Book Company, 1980.

[19] Bird, W. W., "Large Air Supported Radomes for Satellite Communications Ground Stations," *Proc. OSU-RTD Symp. on Electromagnetic Windows*, Ohio State University, 1964.

[20] Bono, R., "Geodesic Domes," Internet page located at http://www.cris.com/~rjbono/html/domes.html, 1996.

[21] D'Amato, R., "Metal Space Frame Radome Design," *Int. Symp. on Structures Technology for Large Radio and Radar Telescope Systems*, Massachusetts Institute of Technology, October 1967.

[22] Kay, A. L., "Electrical Design of Metal Space Frame Radomes," *IEEE Trans. Antennas Propagation*, Vol. AP-13, March 1965.

[23] "Dielectric Space Frame (DSF) Radome Advantages," Antennas for Communication (AFC) Company Internet page located at http://www.ocala.com/afc/imp_mat.gif, 1996.

[24] Rudge, A. W., K. Milne, A. D. Oliver, and P. Knight, *A Handbook of Antenna Design*, London: Peter Peregrinus Ltd., 1986.

[25] Tang, R., and R. W. Burns, "Phased Arrays," Chap. 20 in *Antenna Engineering Handbook,* 3rd ed., R. C. Johnson, Ed., New York: McGraw Hill Book Company, 1993.

[26] Bayliss, E. T., "Design of Monopulse Difference Patterns with Low Sidelobes," *Bell Syst. Tech. J.,* May–June 1968, pp. 623–650.

Selected Bibliography

Bird, W. W., "Large Air Supported Radomes for Satellite Communications Ground Stations," *Proc. OSU-RTD Symp. on Electromagnetic Windows,* Ohio State University, 1964.

Schuchardt, J. M., et al., "Automated Radomc Performance Evaluation in the RFSS Facility at MICOM," *Proc. 15th Symp. on Electromagnetic Windows,* Georgia Institute of Technology, Atlanta, GA, June 1980.

Webb, L. L., "Analysis of Field-of-View versus Accuracy for a Microwave Monopulse," *IEEE Southeascon Proc.,* April 1973, pp. 63–66.

Appendix A

Vector Operations in Various Coordinate Systems

RECTANGULAR COORDINATES

$$\nabla f = \frac{\partial f}{\partial x}\bar{x} + \frac{\partial f}{dy}\bar{y} + \frac{\partial f}{\partial z}\bar{z}$$

$$\nabla \cdot A = \frac{\partial A_x}{\partial x} + \frac{\partial A_y}{\partial y} + \frac{\partial A_z}{\partial z}$$

$$\nabla \times A = \left(\frac{\partial A_z}{\partial y} - \frac{\partial A_y}{\partial z}\right)\bar{x} + \left(\frac{\partial A_x}{\partial z} - \frac{\partial A_z}{\partial x}\right)\bar{y} + \left(\frac{\partial A_y}{\partial x} + \frac{\partial A_x}{\partial y}\right)\bar{z} = \begin{bmatrix} \bar{x} & \bar{y} & \bar{z} \\ \frac{\partial}{\partial x} & \frac{\partial}{\partial y} & \frac{\partial}{\partial z} \\ A_x & A_y & Z_z \end{bmatrix}$$

CYLINDRICAL COORDINATES

$$\nabla f = \frac{\partial f}{\partial r}\bar{r} + \frac{1}{r}\frac{\partial f}{\partial \phi}\bar{\phi} + \frac{\partial f}{\partial z}\bar{z}$$

$$\nabla \cdot A = \frac{1}{r}\frac{\partial}{\partial r} rA_r + \frac{1}{r}\frac{\partial A_\phi}{\partial \phi} + \frac{\partial A_z}{\partial z}$$

$$\nabla \times A = \left(\frac{1}{r}\frac{\partial A_z}{\partial \phi} - \frac{\partial A_\phi}{\partial z}\right)\bar{r} + \left(\frac{\partial A_r}{\partial z} - \frac{\partial A_z}{\partial r}\right)\bar{\phi} + \frac{1}{r}\left(\frac{\partial}{\partial r} rA_\phi - \frac{\partial A_r}{\partial \phi}\right)\bar{z}$$

$$= \begin{bmatrix} \bar{r}\frac{1}{r} & \bar{\phi} & \bar{z}\frac{1}{r} \\ \frac{\partial}{\partial r} & \frac{\partial}{\partial \phi} & \frac{\partial}{\partial z} \\ A_r & rA_\phi & A_z \end{bmatrix}$$

SPHERICAL COORDINATES

$$\nabla f = \frac{\partial f}{\partial r}\bar{r} + \frac{1}{r}\frac{\partial f}{\partial \theta}\bar{\theta} + \frac{1}{r\sin\theta}\frac{\partial f}{\partial \phi}\bar{\phi}$$

$$\nabla \cdot A = \frac{1}{r^2}\frac{\partial}{\partial r} r^2 A_r + \frac{1}{r\sin\theta}\frac{\partial}{\partial \theta}(A_\theta \sin\theta) + \frac{1}{r\sin\theta}\frac{\partial A_\phi}{\partial \phi}$$

$$\nabla \times A = \frac{1}{r\sin\theta}\left[\frac{\partial}{\partial \theta}(A_\phi \sin\theta) - \frac{\partial A_\theta}{\partial \phi}\right]\bar{r} + \frac{1}{r}\left(\frac{1}{\sin\theta}\frac{\partial A_r}{\partial \phi} - \frac{\partial}{\partial r} rA_\phi\right)\bar{\theta} + \frac{1}{r}\left(\frac{\partial}{\partial r} rA_\theta - \frac{\partial A_r}{\partial \theta}\right)\bar{\phi}$$

Appendix B

Propagation Constant and Wave Impedance for Arbitrary Media

In Chapter 5 we presented a boundary value problem based matrix solution for computing the electromagnetic wave transmission and insertion phase delay for a signal propagating through a multilayer dielectric radome wall. That analysis accounted for the dielectric losses in each of the dielectric layers via a loss tangent term associated with the layer. In this appendix, we consider a dielectric material that has both a loss tangent term and a conductivity term, and the electromagnetic wave behavior in such materials. As a result of the mathematical development, the reader should be able to determine the propagation constant, apparent permittivity, and wave impedance.

The assumptions made in the mathematical development that follows apply to most candidate radome materials:

- The material is nonmagnetic, therefore $\mu = \mu_o$.
- The conductivity term is totally real, that is, $\sigma = \sigma'$
- The dielectric term is both real and imaginary, and can be modeled via $\epsilon = \epsilon' - j\epsilon''$

Wave Components in Media

We start with the time-domain form of Maxwell's curl equations:

$$\nabla \times \bar{E} = -\mu_o \frac{\partial \bar{H}}{\partial t} \qquad \text{(B.1)}$$

$$\nabla \times \bar{H} = \sigma \bar{E} + \epsilon \frac{\partial \bar{E}}{\partial t} \qquad \text{(B.2)}$$

The general curl of a vector $\overline{A}$ in rectangular coordinates is:

$$\nabla \times \overline{A} = \left(\frac{\partial A_z}{\partial y} - \frac{\partial A_y}{\partial z}\right)\overline{x} + \left(\frac{\partial A_x}{\partial z} - \frac{\partial A_z}{\partial x}\right)\overline{y} + \left(\frac{\partial Ay}{\partial x} - \frac{\partial A_x}{\partial y}\right)\overline{z} \tag{B.3}$$

In the following development, we will also assume that we have a transverse electromagnetic (TEM) wave traveling in the positive z direction that has only E_x and H_y components. (For normal incidence on a dielectric surface, the reader can derive an identical result by considering a wave that has only E_y and H_x components.) Accordingly, we can write

$$\nabla \times \overline{E} = \frac{\partial E_x}{\partial z}\overline{y} - \frac{\partial E_x}{\partial y}\overline{z} \tag{B.4}$$

$$\nabla \times \overline{H} = \frac{-\partial H_y}{\partial z}\overline{x} + \frac{\partial H_y}{\partial x}\overline{z} \tag{B.5}$$

If the function is both y and x invariant, then these two expressions reduce to:

$$\nabla \times \overline{E} = \frac{\partial E_x}{\partial z}\overline{y} = -\mu_o \frac{\partial H_y}{\partial t}\overline{y} \tag{B.6}$$

or

$$\frac{\partial E_x}{\partial z} = -\mu_o \frac{\partial H_y}{\partial t} \tag{B.7}$$

Similarly,

$$\nabla \times \overline{H} = -\frac{\partial H_y}{\partial z}\overline{x} = \sigma E_x \overline{x} + \epsilon \frac{\partial E_x}{\partial t}\overline{x} \tag{B.8}$$

or

$$-\frac{\partial H_y}{\partial z} = \sigma E_x + \epsilon \frac{\partial E_x}{\partial t} \tag{B.9}$$

For sinusoidal ($e^{j\omega t}$) time variation, (B.7) and (B.9) reduce to

$$\frac{\partial E_x}{\partial z} = -j\omega\mu_o H_y \tag{B.10}$$

$$\frac{\partial H_y}{\partial z} = (\sigma + j\omega\epsilon) E_x \tag{B.11}$$

Now since we propagate as

$$E_x \approx e^{-j\gamma z}; \quad H_y \approx e^{-j\gamma z} \tag{B.12}$$

then

$$\frac{\partial E_x}{\partial z} = -j\gamma E_x \tag{B.13}$$

$$\frac{\partial H_y}{\partial z} = -j\gamma H_y \tag{B.14}$$

Equations (B.10) and (B.11) can be expressed via

$$j\gamma E_x = -j\omega\mu_o H_y \tag{B.15}$$

$$j\gamma H_y = (\sigma + j\omega\epsilon) E_x \tag{B.16}$$

From these last two equations,

$$\frac{E_x}{H_y} = \frac{\omega\mu_o}{\gamma} = \frac{-j\gamma}{(\sigma + j\omega\epsilon)} \tag{B.17}$$

Derivation of Propagation Constant and Apparent Permittivity

The propagation constant can be derived from (B.17):

$$\gamma^2 = \omega^2\mu\epsilon - j\omega\mu_o\sigma \tag{B.18}$$

Or, in terms of the real and imaginary parts of the permittivity:

$$\gamma^2 = \omega^2\mu_o\epsilon' - j\omega^2\mu_o\epsilon'' - j\omega\mu_o\sigma = \omega^2\mu_o\epsilon'\left[1 - j\left(\frac{\epsilon''}{\epsilon'} + \frac{\sigma}{\omega\epsilon'}\right)\right] \tag{B.19}$$

From this last equation, together with the definition of loss tangent, we obtain the result:

$$\gamma = \sqrt{\omega^2\mu_o\epsilon'}\sqrt{1 - j\left(\tan\delta + \frac{\sigma}{\omega\epsilon'}\right)} = k\sqrt{1 - j\left(\tan\delta + \frac{\sigma}{\omega\epsilon'}\right)} \tag{B.20}$$

Comparing this to the expression in Chapter 5 for a material without a conductivity term, we obtain the following for a complex apparent permittivity:

$$\epsilon = \epsilon'\left[1 - j\left(\tan\delta + \frac{\sigma}{\omega\epsilon'}\right)\right] \tag{B.21}$$

Or, the relative apparent permittivity is

$$\epsilon_r = \epsilon_r'\left[1 - j\left(\tan\delta + \frac{\sigma}{\omega\epsilon_o\epsilon_r'}\right)\right] \tag{B.22}$$

Wave Impedance

From (B.20), we can obtain

$$\gamma = k\left[1+\left(\tan\delta+\frac{\sigma}{\omega\epsilon'}\right)^2\right]^{0.25} e^{-j0.5\tan^{-1}\left(\tan\delta+\frac{\sigma}{\omega\epsilon'}\right)} \tag{B.23}$$

If we incorporate this into (B.17) and express the result in terms of the relative apparent permittivity, we obtain the wave impedance for normal incidence:

$$Z = \frac{\frac{120\pi}{\sqrt{\epsilon'}} e^{j0.5\tan^{-1}\left(\tan\delta+\frac{\sigma}{\omega\epsilon_o\epsilon_r'}\right)}}{\left[1+\left(\tan\delta+\frac{\sigma}{\omega\epsilon_o\epsilon_r'}\right)^2\right]^{0.25}} \tag{B.24}$$

This can be used to calculate the interfacial Fresnel reflection and transmission coefficients.

For other than normal incidence, assume the wave is propagating from one medium to a second medium. The impedance just inside the interfacial boundary in the direction of propagation is modified for perpendicular and parallel polarization via the following approximations. For perpendicular polarization:

$$Z_{\perp} = \frac{120\pi\cos\theta}{\sqrt{\epsilon_r - \sin^2\theta}} \tag{B.25}$$

and for parallel polarization:

$$Z_{\parallel} = \frac{120\pi\sqrt{\epsilon_r - \sin^2\theta}}{\epsilon_r\cos\theta} \tag{B.26}$$

where θ is the angle of incidence with respect to the surface normal, and the reader should use the apparent relative permittivity given by (B.22).

Appendix C

Radiation From Elemental Currents

The purpose of this appendix is to derive expressions for both the electromagnetic field and the array factor for an ensemble of infinitesimal (dipole) current elements. Three cases are considered as illustrated in Figure C.1:

1. Current elements that are z-directed;
2. Current elements that are x-directed;
3. Current elements that are y-directed.

The far-field radiation pattern of an infinitesimal dipole element follows from solution of Maxwell's equations. Assuming a time harmonic field that is $e^{j\omega t}$ dependent, the fundamental relations are expressed as the curl and divergence operators:

$$\nabla \times \overline{H} = j\omega\epsilon_o \overline{E} + \overline{J} \tag{C.1}$$

$$\nabla \times \overline{E} = j\omega\mu_o \overline{H} \tag{C.2}$$

where ϵ_o and μ_o are the permittivity and permeability of free space, $\overline{B} = \mu_o \overline{H}$ and $\overline{D} = \epsilon_o \overline{E}$.

Now, define a magnetic vector potential via

$$\overline{H} = \nabla \times \overline{A} \tag{C.3}$$

then from (C.2) and (C.3), we obtain

$$\nabla \times (\overline{E} + j\omega\mu_o \overline{A}) = 0 \tag{C.4}$$

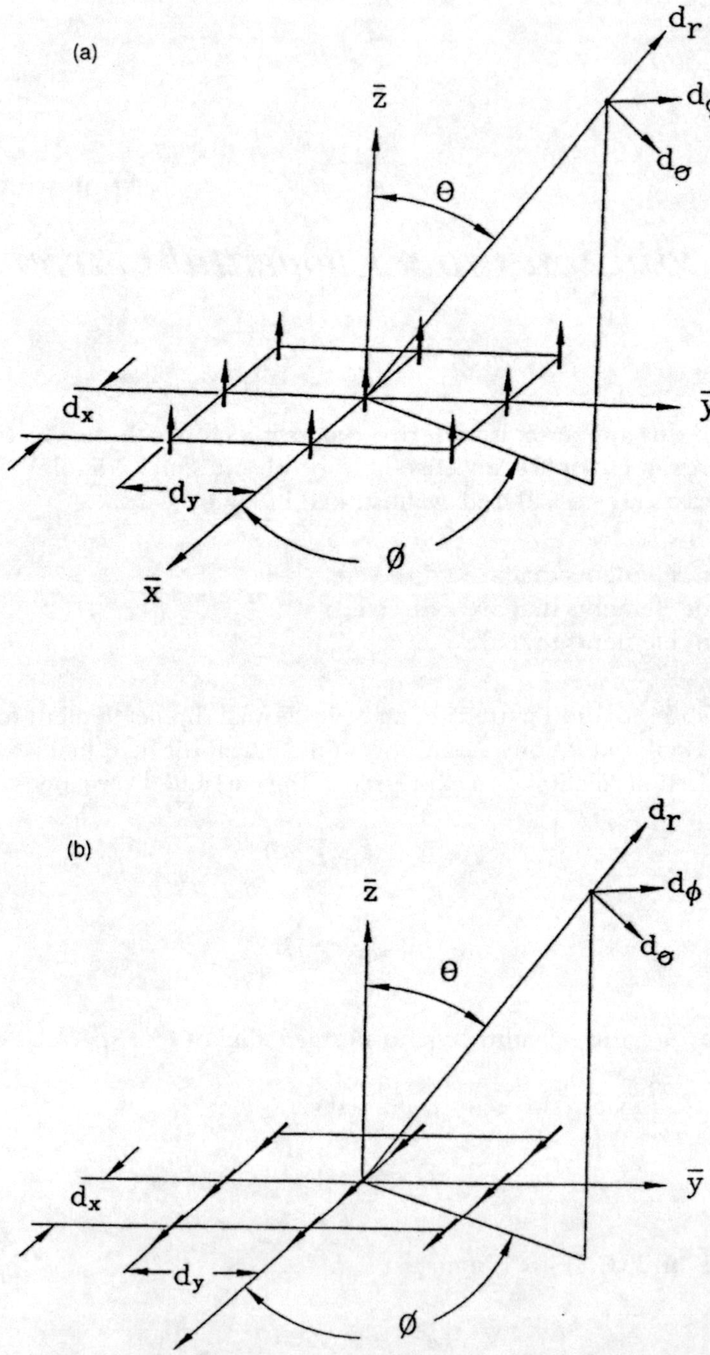

Figure C.1 Radiation from infinitesimal currents: (a) z-directed, (b) x-directed, and (c) y-directed.

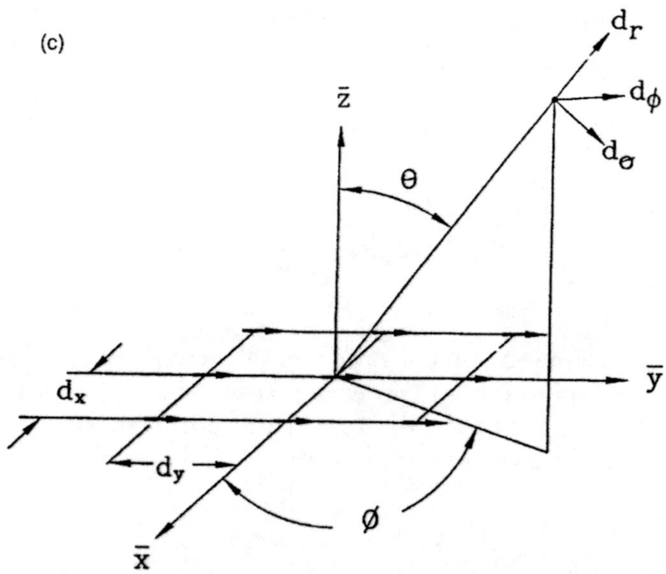

Figure C.1 (continued)

Because any curl-free vector can be defined as the gradient of a scalar function, one can define an electric scalar potential Φ where

$$\overline{E} + j\omega\mu_o\overline{A} = -\nabla\Phi \tag{C.5}$$

From (C.1), (C.3), and (C.5), we obtain

$$\nabla(\nabla \cdot \overline{A}) - \nabla^2\overline{A} - k^2\overline{A} = -j\omega\epsilon_o\nabla\Phi + \overline{J} \tag{C.6}$$

A vector is uniquely specified by both the curl and divergence. If we choose

$$\nabla(\nabla \cdot \overline{A}) = -j\omega\epsilon_o\nabla\Phi \tag{C.7}$$

$$\nabla^2\overline{A} + k^2\overline{A} = -\overline{J} \tag{C.8}$$

we obtain the well-known Helmholtz wave equation. Solutions to the field components for $\overline{A}$ follow from (C.3), (C.5), and (C.7):

$$\overline{H} = \nabla \times \overline{A} \tag{C.9}$$

$$\overline{E} = -j\omega\mu_o \overline{A} + \frac{1}{j\omega\epsilon_o}\nabla(\nabla \cdot \overline{A}) \tag{C.10}$$

Because rectangular components of $\overline{A}$ only have corresponding components of $\overline{J}$ as their source, the technique is first developed by assuming only a *z*-directed current element at the origin of a standard spherical coordinate system, as depicted in Figure C.1(a). We assume current of magnitude I_o and length *d*.

$$\nabla^2 A_z - k^2 A_z = I_o d \tag{C.11}$$

where *k* is the free-space wave number, *h* is the infinitesimal dipole length, and I_o is the current magnitude. Assume A_z is *r* dependent. The differential equation to be solved is

$$\frac{1}{r^2 \partial r}\frac{\partial}{\partial r}\left(r^2 \frac{\partial A_z}{\partial r}\right) + k^2 A_z = I_o d \tag{C.12}$$

$$A_z = I_o \frac{e^{-jkr}}{4\pi r} \tag{C.13}$$

From (C.9)

$$\overline{H} = \nabla \times \bar{z} A_z = \frac{\partial A}{\partial r}\bar{r} \times \bar{z} \tag{C.14}$$

from which we obtain (in the far field)

$$H_\phi = jkI_o d\frac{e^{-jkr}}{4\pi r}\sin\theta \tag{C.15}$$

$$E_\theta = jk\eta I_o d\frac{-jkr}{4\pi r}\sin\theta \tag{C.16}$$

where

$$\eta = \sqrt{\frac{\mu_o}{\epsilon_o}} = 120\pi \tag{C.17}$$

Equations (C.15) and (C.16) can be shown to be equivalent to

$$E_\theta = j\omega\epsilon_o \sin\theta A_z \tag{C.18}$$

$$H_\phi = \frac{1}{\eta} E_\theta \tag{C.19}$$

The relationship between far-field patterns and aperture currents follows from linear superposition of the radiated fields from each of the infinitesimal currents. For example, consider the *z*-directed current distribution shown in Figure C.1(a). The location of currents in the *x* direction is

$$x = md_x \tag{C.20}$$

and the location of currents in the *y* direction is

$$y = nd_y \tag{C.21}$$

In these last two equations, m and n are integers, where $m = 0, 1, 2, \ldots, M$ and $n = 0, 1, 2, \ldots, N$, and d_x and d_y are the current spacings in the x and y directions, respectively. The total current is given by the formula

$$I = \sum_{m=0}^{M} \sum_{n=0}^{N} I_{mn} \delta(x - md_x)\,\delta(y - nd_y)\,\delta(z) \tag{C.22}$$

where the Dirac delta functions are defined by the following relations:

$$\delta(p) \begin{cases} 0; & \text{for all } p \neq 0 \quad \text{(C.23)} \\ 1; & \text{for } p = 0 \quad \text{(C.24)} \end{cases}$$

Using this discrete current distribution in (C.13), we obtain

$$A_z = \iiint \frac{e^{-jkr}}{4\pi r} \delta(z) \sum_{m=0}^{M} \sum_{n=0}^{N} I_{mn} \delta(x - md_x)\, \delta(y - nd_y)\, dx\, dy\, dz \tag{C.25}$$

For a fixed distance from the antenna, r_o, this reduces to

$$A_z = \left[\frac{e^{-jkr_o}}{4\pi r_o} \right] \sum_{m=0}^{M} \sum_{n=0}^{N} I_{mn} e^{-jk\sin\theta(md_x \cos\phi + nd_y \sin\phi)} \tag{C.26}$$

This result is sufficient to obtain far-field radiation patterns. Specifically, the far-field pattern is a simple sum of the current contributions of the ensembles of infinitesimal current sources, that is, by incorporation of this expression into (C.18):

$$E_\theta = \sin(\theta) \left[\frac{e^{-jkr_o}}{4\pi r_o} \right] \sum_{m=0}^{M} \sum_{n=0}^{N} I_{mn} e^{j\psi_{mn}} \tag{C.27}$$

where the space phase term is defined via

$$\psi_{mn} = k \sin\theta (md_x \cos\phi + nd_y \sin\phi) \tag{C.28}$$

In (C.27) we can drop the term in brackets [ ] because it is a scale constant. If we do so, we obtain the following array factor:

$$E_\theta^{\mathrm{AF}} = \sin(\theta) \sum_{m=0}^{M} \sum_{n=0}^{N} I_{mn} e^{j\psi_{mn}} \tag{C.29}$$

If the currents are *x*-directed, a similar development leads to the following expression for the array factor for the field component aligned with the *x*-polarization axis:

$$E_x^{\mathrm{AF}} = \cos \left[\tan^{-1} \left(\frac{\sin\theta \cos\phi}{\sqrt{\cos^2\theta + \sin^2\theta \sin^2\phi}} \right) \right] \sum_{m=0}^{M} \sum_{n=0}^{N} I_{mn} e^{j\psi_{mn}} \tag{C.30}$$

If the currents are *y*-directed, the array factor for the field component aligned with the *y*-polarization axis is

$$E_y^{\mathrm{AF}} = \cos\left[\tan^{-1}\left(\frac{\sin\theta\sin\phi}{\sqrt{\cos^2\theta+\sin^2\theta\sin^2\phi}}\right)\right]\sum_{m=0}^{M}\sum_{n=0}^{N} I_{mn}e^{j\psi_{mn}} \tag{C.31}$$

Appendix D

Listing of Program Tangent Ogive Radome

VARIABLE LIST

L0	radome length (inches)
D0	radome diameter (inches)
R,B	parameters used to describe tangent ogive geometry (inches)
xg	x coordinate of gimbal point (inches)
yg	y coordinate of gimbal point (inches)
zg	z coordinate of gimbal point (inches)
delta.a	gimbal axis offset (inches)
delta.b	antenna face offset from gimbal axis (inches)
A.dia	antenna diameter (inches)
A.rad	antenna radius (inches)
xi,yi,zi	a general point along the line between the initial point and the intercept point (inches)
xp,yp,zp	intercept point on the radome surface (inches)
kx,ky,kz	unit propagation vector (inches ^ -1)
Ci	convergence constant (dimensionless)
nx,ny,nz	unit vector normal to the surface at intercept point (inches)
theta	angle between the propagation vector and the normal vector (radians)
ERFF	computational error function (inches)
freq	frequency (GHz)
LAM	wavelength (inches)
k0	wavenumber (inches ^ -1)
'N.lay	number of layers (dimensionless)
er(I)	relative permittivity of the *I*th layer (dimensionless)
thk(I)	thickness of the *I*th layer (inches)

ltan(I)	loss tangent of the *I*th layer (dimensionless)
COPOL	copolarized radome loss (decibels)
XPOL	crosspolarized radome loss (decibels)
AZBSE	azimuth channel boresight error (milliradians)
ELBSE	elevation channel boresight error (milliradians)
AZ,EL	azimuth and elevation angles (degrees)
ANG.STEP	computational angular steps (degrees)
ANG.MAX	maximum gimbal scan (degrees)

```
* * *  Automatic data file generation for plotting  * * *

CLEAR
RESET

REDIM thk(15), er(15), ltan(15), R(16), U2(16), U3(16), z(16)
REDIM AR(5), BR(5), CR(5), AI(5), BI(5), Ci(5)
REDIM K(5), N(5), TPERP(50), PS(50)
REDIM RAZP(65, 65), RELP(65, 65), IELP(65, 65), IAZP(65, 65)
REDIM XRAZP(65, 65), XRELP(65, 65), XIELP(65, 65), XIAZP(65, 65)

- - - - - - - - - - - - - - - - - - - - - - - - - - - - - - - - - - - - - -
TYPE variables
  EL AS INTEGER
  AZ AS INTEGER
  IIY AS INTEGER
  IIZ AS INTEGER
  IY AS INTEGER
  IZ AS INTEGER
  NPY AS INTEGER
  NPZ AS INTEGER
  IMY AS INTEGER
  IMZ AS INTEGER
  I AS INTEGER
  J AS INTEGER
  NA AS INTEGER
  NB AS INTEGER
  MO AS INTEGER
END TYPE
COLOR 15, 1: CLS

'- - - - - - - - - - - - - - - - - - - - - - - - - - - - - - - - - - - - -
     INPUT "Do you want to send data output to a printer? : ", print.data$
     INPUT "Enter Number of Layers : ", N.lay
     FOR I = 1 TO N.lay
     PRINT "For layer number :  "; I
     INPUT "Enter thickness (inches)  ; ", thk(I)
     INPUT "Enter relative permittivity : ", er(I)
     INPUT "Enter loss tangent : ", ltan(I)
     NEXT I
```

```
     IF print.data$ = "Y" OR print.data$ = "y" THEN
     LPRINT "Enter Number of Layers : "; N.lay
     FOR I = 1 TO N.lay
     LPRINT "For layer number :  "; I
     LPRINT "thickness (inches)  ; "; thk(I)
     LPRINT "relative permittivity : "; er(I)
     LPRINT "loss tangent : "; ltan(I)
     NEXT I
     LPRINT "": LPRINT ""
     END IF
- - - - - - - - - - - - - - - - - - - - - - - - - - - - - - - - - - - - -
     PI = 3.14159265#
     RAD = PI / 180

     PRINT " "
    INPUT "Enter antenna diameter (inches) : ", A.dia
       A.rad = A.dia / 2
    INPUT "Enter x-coordinate of gimbal location (inches) : ", xg
    INPUT "Enter y-coordinate of gimbal location (inches) : ", yg
    INPUT "Enter z-coordinate of gimbal location (inches) : ", zg
    INPUT "Enter delta-a gimbal offset (inches) : ", delta.a
    INPUT "Enter delta-b gimbal offset (inches) : ", delta.b
    INPUT "For AZ/EL gimbal input(0), for EL/AZ gimbal input(1) : ", GIMBAL
    PRINT "": PRINT ""
    INPUT "Enter maximum gimbal computation angle (degrees) : ", ANG.MAX
    INPUT "Enter gimbal computation angle steps (degrees) : ", ANG.STEP
    INPUT "Enter radome length (inches) : ", L0
    INPUT "Enter radome diameter (inches) : ", D0

IF print.data$ = "Y" OR print.data$ = "y" THEN
       LPRINT "antenna diameter (inches) : "; A.dia
       LPRINT "x-coordinate of gimbal location (inches) : "; xg
       LPRINT "y-coordinate of gimbal location (inches) : "; yg
       LPRINT "z-coordinate of gimbal location (inches) : "; zg
       LPRINT "delta-a gimbal offset (inches) : "; delta.a
       LPRINT "delta-b gimbal offset (inches) : "; delta.b
       IF GIMBAL = 0 THEN LPRINT "For AZ/EL gimbal case"
       IF GIMBAL = 1 THEN LPRINT "For EL/AZ gimbal case"
       LPRINT "radome length (inches) : "; L0
       LPRINT "radome diameter (inches) : "; D0
       LPRINT "": LPRINT ""
       END IF

'      Works for L0/D0=0.5, or higher, fineness ratios
'      One time calculation for radome
       B = (4 * L0 ^ 2 - D0 ^ 2) / (4 * D0)
       R = (4 * L0 ^ 2 + D0 ^ 2) / (4 * D0)
       IF B = 0 THEN B = 1E-12
- - - - - - - - - - - - - - - - - - - - - - - - - - - - - - - - - - - - -
       PRINT "": PRINT ""
       INPUT "Enter computation frequency (GHz) : ", freq
```

```
        PRINT "": PRINT ""
        LAM = 11.81 / freq
        K0 = .532015 * freq

INPUT "Do you want to create a delimited ASCII data output file? : ",
 create.file$
redo.filename:
    IF create.file$ = "Y" OR create.file$ = "y" THEN
    INPUT "enter filename for data output file (no extension) : ",
 filename$
    IF INSTR(filename$, ".") > 0 THEN BEEP: GOTO redo.filename
      ELSE
    filename$ = "DATAFILE"   ' <-- don't use extension on filename
    END IF

'       SET POLARIZATION:

        FOR MO = 0 TO 1
        IF MO = 0 THEN polarization$ = "V"
        IF MO = 1 THEN polarization$ = "H"

        IF polarization$ = "V" THEN
                EAZ = 0: EEL = 1
        END IF

        IF polarization$ = "H" THEN
                EAZ = 1: EEL = 0
        END IF


PRINT "FREQUENCY (GHZ) =    "; freq; "   POLARIZATION =    "; polarization$
PRINT "": PRINT ""

IF print.data$ = "Y" OR print.data$ = "y" THEN
LPRINT "FREQUENCY (GHZ) =    "; freq; "   polarization =    "; polarization$
LPRINT "": LPRINT "": LPRINT ""
END IF

'       ARP is distance from antenna to ref. plane (inches)
        ARP = L - xg + 3

        first.time$ = "NO"

'       INTEGRATION INDEX IN REFERENCE PLANE
        NPZ = 1: NPY = 1
'       Grid spacing in reference plane in wavelengths:
        RPSP = .75
'       Reference plane diameter:
        WSEP = RPSP * LAM
        RPDIA = 2 * (NPY - 1) * WSEP

'       Withradome$ = "NO" is first time through computation of SREF
```

```
'              (normalization for transmission coefficient)

        AZ = 0: EL = 0
        AZOFFSET = 0: ELOFFSET = 0
        withradome$ = "NO": Wantgainref$ = "YES"
        GOSUB INTEGRATION
        Wantgainref$ = "NO"
        GOSUB MONOPULSE
        SUMMAX = SUMTOT

'       CALCULATION OF MONOPULSE ERROR CHANNEL SENSITIVITIES:
        AZ = 0: EL = 0
        AZOFFSET = .05: ELOFFSET = 0
        GOSUB MONOPULSE
        KAZ = ABS(VAZ) / AZOFFSET

        AZ = 0: EL = 0
        AZOFFSET = 0: ELOFFSET = .05
        GOSUB MONOPULSE
        KEL = ABS(VEL) / ELOFFSET

        IF KAZ = 0 THEN KAZ = .000001: IF KEL = 0 THEN KEL = .000001

        AZOFFSET = 0: ELOFFSET = 0

        withradome$ = "YES"

'- - - - - - SCANNER FOR BSE COMPUTATIONS - - - - - -
'- - - - - - load arrays  - - - - - - - - - - - - - -

        AZOFFSET = 0: ELOFFSET = 0
        EL = 0
        FOR AZ = 0 TO ANG.MAX STEP ANG.STEP

        in$ = INKEY$: IF in$ = CHR$(27) THEN END

PRINT USING " Calculating For:  Elevation =+###  Azimuth =+###"; EL, AZ

IF print.data$ = "Y" OR print.data$ = "y" THEN
LPRINT USING " Calculating For:  Elevation =+###  Azimuth =+###"; EL, AZ
END IF



GOSUB MONOPULSE

        IF SUMTOT = 0 THEN SUMTOT = 10 ^ -6
        ldb = 20 * LOG(SUMTOT / SUMMAX) / LOG(10)
        IF XSUMTOT = 0 THEN XSUMTOT = .000001
        xldb = 20 * LOG(XSUMTOT / SUMMAX) / LOG(10)
        xldb = xldb + ldb
```

```
        azbse = ABS(17.4533 * VAZ / KAZ)
        IF AZ < 0 THEN azbse = -ABS(azbse)

        elbse = ABS(17.4533 * VEL / KEL)
        IF EL < 0 THEN elbse = -ABS(elbse)

        PRINT " "
        PRINT "   COPOL(dB)     ="; ldb
        PRINT "   XPOL(dB)      ="; xldb
        PRINT "   AZBSE(mrad)  ="; azbse
        PRINT "   ELBSE(mrad)    ="; elbse
        PRINT "": PRINT ""

        IF print.data$ = "Y" OR print.data$ = "y" THEN
        LPRINT " "
        LPRINT "   COPOL(dB)     ="; ldb
        LPRINT "   XPOL(dB)      ="; xldb
        LPRINT "   AZBSE(mrad)  ="; azbse
        LPRINT "   ELBSE(mrad)    ="; elbse
        LPRINT "": PRINT ""
        END IF

     NEXT AZ


        AZ = 0
        FOR EL = 0 TO ANG.MAX STEP ANG.STEP

        in$ = INKEY$: IF in$ = CHR$(27) THEN END

PRINT USING " Calculating For:  Elevation =+###  Azimuth =+###"; EL, AZ

IF print.data$ = "Y" OR print.data$ = "y" THEN
LPRINT USING " Calculating For:  Elevation =+###  Azimuth =+###"; EL, AZ
END IF

GOSUB MONOPULSE

        IF SUMTOT = 0 THEN SUMTOT = 10 ^ -6
        ldb = 20 * LOG(SUMTOT / SUMMAX) / LOG(10)
        IF XSUMTOT = 0 THEN XSUMTOT = .000001
        xldb = 20 * LOG(XSUMTOT / SUMMAX) / LOG(10)
        xldb = xldb + ldb

        azbse = ABS(17.4533 * VAZ / KAZ)
        IF AZ < 0 THEN azbse = -ABS(azbse)

        elbse = ABS(17.4533 * VEL / KEL)
        IF EL < 0 THEN elbse = -ABS(elbse)

        PRINT " "
```

```
        PRINT "    COPOL(dB)     ="; ldb
        PRINT "    XPOL(dB)      ="; xldb
        PRINT "    AZBSE(mrad)  ="; azbse
        PRINT "    ELBSE(mrad)     ="; elbse
        PRINT "": PRINT ""

        IF print.data$ = "Y" OR print.data$ = "y" THEN
        LPRINT " "
        LPRINT "    COPOL(dB)     ="; ldb
        LPRINT "    XPOL(dB)      ="; xldb
        LPRINT "    AZBSE(mrad)  ="; azbse
        LPRINT "    ELBSE(mrad)     ="; elbse
        LPRINT "": PRINT ""
        END IF

        NEXT EL

NEXT MO

END
```

MONOPULSE

Subroutine Monopulse Antenna

Forms an array superimposed onto antenna.diameter, with respective sine and cosine weighting functions:

```
        SUMRE = 0: SUMIM = 0: DAZRE = 0: DAZIM = 0
        DELRE = 0: DELIM = 0: XSUMRE = 0: XSUMIM = 0

        FOR JZ = 0 TO 16
        FOR JY = 0 TO 16

        YA = ((-8 + JY) / 16) * A.dia
        ZA = ((-8 + JZ) / 16) * A.dia

        IF SQR(YA ^ 2 + ZA ^ 2) > A.rad THEN GOTO 50

        GOSUB INTEGRATION

        ' Assume cosine illumination:
weight = ABS(COS(YA * PI / A.dia) * COS(ZA * PI / A.dia))

        SUMRE = (RELP(JY, JZ) + RAZP(JY, JZ)) * weight + SUMRE
        SUMIM = (IELP(JY, JZ) + IAZP(JY, JZ)) * weight + SUMIM

        DAZRE = (RELP(JY, JZ) + RAZP(JY, JZ)) * weight * SGN(ZA) + DAZRE
```

```
        DAZIM = (IELP(JY, JZ) + IAZP(JY, JZ)) * weight * SGN(ZA) + DAZIM

        DELRE = (RELP(JY, JZ) + RAZP(JY, JZ)) * weight * SGN(YA) + DELRE
        DELIM = (IELP(JY, JZ) + IAZP(JY, JZ)) * weight * SGN(YA) + DELIM

        XSUMRE = (XRELP(JY, JZ) + XRAZP(JY, JZ)) * weight + XSUMRE
        XSUMIM = (XIELP(JY, JZ) + XIAZP(JY, JZ)) * weight + XSUMIM
50   '- - - - CONTINUE
NEXT JY
NEXT JZ

'- - - - Fix to prevent overflows

        IF SUMRE = 0 THEN SUMRE = .000001
        IF DAZRE = 0 THEN DAZRE = .000001
        IF DELRE = 0 THEN DELRE = .000001
        IF XSUMRE = 0 THEN XSUMRE = .000001#

        SUMANG = ATN(SUMIM / SUMRE): DAZANG = ATN(DAZIM / DAZRE)
        DELANG = ATN(DELIM / DELRE)

 SUMTOT = SQR(SUMRE ^ 2 + SUMIM ^ 2): XSUMTOT = SQR(XSUMRE ^ 2 + XSUMIM ^
 2)
  VAZ = (SQR(DAZRE ^ 2 + DAZIM ^ 2) / SUMTOT) * SGN(COS(DAZANG - SUMANG))
'   VEL = (SQR(DELRE ^ 2 + DELIM ^ 2) / SUMTOT) * SGN(COS(DELANG - SUMANG))
   VEL = (SQR(DELRE ^ 2 + DELIM ^ 2) / SUMTOT)
RETURN

'       SUBROUTINE INTEGRATION:
INTEGRATION:

RAZP(JY, JZ) = 0: RELP(JY, JZ) = 0: IELP(JY, JZ) = 0: IAZP(JY, JZ) = 0
XRAZP(JY, JZ) = 0: XRELP(JY, JZ) = 0: XIELP(JY, JZ) = 0: XIAZP(JY, JZ) = 0

        CAZ = COS(AZOFFSET * RAD): CEL = COS(ELOFFSET * RAD)
        SAZ = SIN(AZOFFSET * RAD): SEL = SIN(ELOFFSET * RAD)

        PX = 0: PY = YA: PZ = ZA
        GOSUB ROTATION
        XAROT = xp: YAROT = yp: ZAROT = zp

'       Note: only XAROT contributes to space phase term

'       INDEX ON REFERENCE PLANE POINTS:

        FOR IY = -NPY TO NPY
        FOR IZ = -NPZ TO NPZ

        XRP = ARP: YRP = WSEP * IY: ZRP = WSEP * IZ

'       Note:  use a circular reference plane:
```

```
        RADIUS = SQR(YRP ^ 2 + ZRP ^ 2)
        IF (RADIUS) > ((RPDIA / 2) + .000001) THEN GOTO 1100

'       ROTATE REFERENCE PLANE POINTS THRU GIMBAL ANGLES

        CAZ = COS(AZ * RAD): CEL = COS(EL * RAD)
        SAZ = SIN(AZ * RAD): SEL = SIN(EL * RAD)
        PX = XRP: PY = YRP: PZ = ZRP
        GOSUB ROTATION
        XRPROT = xp: YRPROT = yp: ZRPROT = zp

'NOTE:  XRPROT, YRPROT, ZRPROT are rotated reference plane points

'       ROTATE ANTENNA PLANE POINTS THRU GIMBAL ANGLES ONLY:

        PX = 0: PY = YA: PZ = ZA
        GOSUB ROTATION

'       XP, YP, ZP are rotated antenna points

'       TO EXPRESS IN TRUE RADOME COORDINATES, APPLY TRANSLATIONS:

 XARC = xp + xg: YARC = yp + yg: ZARC = zp + zg
 XRPRC = XRPROT + xg + xp: YRPRC = YRPROT + yg + yp: ZRPRC = ZRPROT
 + zg + zp

'       DISTANCE FROM ANTENNA POINT TO POINT IN REFERENCE PLANE
        RPDIS = SQR((XRPRC - XARC) ^ 2 + (YRPRC - YARC) ^ 2 + (ZRPRC -
ZARC) ^ 2)

'       UNIT VECTOR FROM ANTENNA POINT TO POINT IN REFERENCE PLANE:
'       K(1),K(2),K(3) are x,y,z components, respectively

        K(1) = (XRPRC - XARC) / RPDIS
        K(2) = (YRPRC - YARC) / RPDIS
        K(3) = (ZRPRC - ZARC) / RPDIS

'       ANTENNA POINT PS(1),PS(2),PS(3) are x,y,z components, respectively
        PS(1) = XARC: PS(2) = YARC: PS(3) = ZARC

'       1st time through computation of SREF (normalization for
trans.coef.)

        IF withradome$ = "NO" THEN GOTO 1001

'       RAY TRACE TO FIND RADOME INTERCEPT POINT, SURFACE NORMAL AT
INTERCEPT  POINT, AND RAY ANGLE WITH RESPECT TO SURFACE NORMAL
VECTOR

          GOSUB RAYINTERCEPT

1001    'CONTINUE
```

```
'       RESOLVE INCIDENT E-FIELD INTO PARA.&PERP.COMPONENTS, ADD EFFECT OF
'       RADOME TRANSMISSION, AND ACCUMULATE THE RESULTS!

        GOSUB DECOMP

1100    NEXT IZ
        NEXT IY

'       COMPUTATION OF ANTENNA TERMINAL VOLTAGE
   IF (Wantgainref$ = "YES") THEN
   SREF = SQR(RELP(JY, JZ) ^ 2 + IELP(JY, JZ) ^ 2 + RAZP(JY, JZ) ^ 2 +
 IAZP(JY, JZ) ^ 2)

        ELSE

   RELP(JY, JZ) = RELP(JY, JZ) / SREF: RAZP(JY, JZ) = RAZP(JY, JZ) / SREF
   IELP(JY, JZ) = IELP(JY, JZ) / SREF: IAZP(JY, JZ) = IAZP(JY, JZ) / SREF
   XRELP(JY, JZ) = XRELP(JY, JZ) / SREF: XRAZP(JY, JZ) = XRAZP(JY, JZ) /
 SREF
   XIELP(JY, JZ) = XIELP(JY, JZ) / SREF: XIAZP(JY, JZ) = XIAZP(JY, JZ) /
 SREF

        END IF

RETURN

'       SUBROUTINE COORDINATE ROTATION:
ROTATION:

        IF GIMBAL = 0 THEN
'       AZ/EL Gimbal modeling

xp = PX * CAZ * CEL - PY * CAZ * SEL - PZ * SAZ
yp = PX * SEL + PY * CEL + 0
zp = PX * SAZ * CEL - PY * SAZ * SEL + PZ * CAZ
        END IF

        IF GIMBAL = 1 THEN
'       EL/AZ Gimbal modeling

 xp = PX * CEL * CAZ - PY * SEL - PZ * CEL * SAZ
 yp = PX * SEL * CAZ + PY * CEL - PZ * SEL * SAZ
 zp = PX * SAZ + 0 + PZ * CAZ
        END IF
RETURN

'       SUBROUTINE GIMBAL ROTATION:
GIMBALROT:
        IF GIMBAL = 0 THEN
'       AZ/EL Gimbal modeling

xp = delta.b * CAZ * CEL - PY * CAZ * SEL - PZ * SAZ
```

```
xp = xp + delta.a * CEL
yp = PX * SEL + PY * CEL + 0
yp = yp + delta.a * SEL
zp = PX * SAZ * CEL - PY * SAZ * SEL + PZ * CAZ
        END IF

        IF GIMBAL = 1 THEN
'       EL/AZ Gimbal modeling

   xp = delta.b * CEL * CAZ - PY * SEL - PZ * CEL * SAZ
   xp = xp + delta.z * CEL
   yp = PX * SEL * CAZ + PY * CEL - PZ * SEL * SAZ
   yp = yp + delta.a * SEL
   zp = PX * SAZ + 0 + PZ * CAZ
        END IF
RETURN

RAYINTERCEPT:
'       PS(1), PS(2), PS(3) are rotated and translated antenna points
'       K(1), K(2), K(3) are ray vector components from antenna points
 toward
'       points in reference plane!

        x0 = PS(1): y0 = PS(2): z0 = PS(3)
        kx = K(1): ky = K(2): kz = K(3)

'       ROOTFINDER TO FIND Ci
'       Increments in inches:
        Ci = 0
        FOR I = 1 TO 1000
        Ci = Ci + 1
        xi = x0 + Ci * kx
        yi = y0 + Ci * ky
        zi = z0 + Ci * kz
        GOSUB 100
        IF ERFF = 0 GOTO 200
        IF ERFF < 0 THEN EXIT FOR
        NEXT I

        FOR I = 1 TO 10
'       Increments in tenths of inches:
        Ci = Ci - .1
        xi = x0 + Ci * kx
        yi = y0 + Ci * ky
        zi = z0 + Ci * kz
        GOSUB 100
        IF ERFF = 0 GOTO 200
        IF ERFF < 0 THEN EXIT FOR
        NEXT I

        FOR I = 1 TO 10
'       Increments in hundreths of inches:
```

```
        Ci = Ci + .01
        xi = x0 + Ci * kx
        yi = y0 + Ci * ky
        zi = z0 + Ci * kz
        GOSUB 100
        IF ERFF = 0 GOTO 200
        IF ERFF < 0 THEN EXIT FOR
        NEXT I

200     'continue

'       Calculate normal vector at intercept point:

        xp = xi: yp = yi: zp = zi
        rp = SQR(zp ^ 2 + yp ^ 2)
        nz = ((rp + B) / SQR((rp + B) ^ 2 + xp ^ 2)) * zp / rp
        ny = ((rp + B) / SQR((rp + B) ^ 2 + xp ^ 2)) * yp / rp
        nx = xp / SQR((rp + B) ^ 2 + xp ^ 2)

        N(1) = nx: N(2) = ny: N(3) = nz
        x1 = xi: y1 = yi: z1 = zi
'- - - - Note: x1, y1, z1 are now the x, y, z intercept points

'- - - - N(1),N(2),N(3) are X,Y,Z components of vector at icpt pt.

'- - - - K(1), K(2), K(3) were x, y, z components of vector at antenna
 point in
'               direction of point in external reference plane!

'- - - - T(1), T(2), T(3) are x, y, z components of perpendicular vector

    TPERP(1) = K(2) * N(3) - K(3) * N(2)
    TPERP(2) = K(3) * N(1) - K(1) * N(3)
    TPERP(3) = K(1) * N(2) - K(2) * N(1)

    U = SQR(TPERP(1) * TPERP(1) + TPERP(2) * TPERP(2) + TPERP(3) *
 TPERP(3))

    TPERP(1) = TPERP(1) / U
    TPERP(2) = TPERP(2) / U
    TPERP(3) = TPERP(3) / U

' TPERP(1), TPERP(2), TPERP(3)
'END

'- - - - WAVE ANGLE WITH RESPECT TO SURFACE NORMAL:
NDOTK = (N(1) * K(1) + N(2) * K(2) + N(3) * K(3))
THETA = ABS(ATN(SQR(1 - NDOTK ^ 2) / NDOTK))

RETURN

100     'Subroutine Tangent Ogive Shape
```

```
        IF zi = 0 AND yi = 0 THEN zi = .000001

        IF xi <= L0 THEN
        ERFF = (SQR(ABS(R ^ 2 - xi ^ 2)) - B) - SQR(zi ^ 2 + yi ^ 2)
        END IF

        IF xi > L0 THEN
        ERFF = -(SQR(ABS(R ^ 2 - (2 * L0 - xi) ^ 2)) - B) - SQR(zi ^ 2 +
 yi ^ 2)
        END IF
        RETURN


'       SUBROUTINE E-FIELD DECOMPOSITION:
DECOMP:

'- - - - Rotate incident wavefront

    PX = 0: PY = 1: PZ = 0

    GOSUB ROTATION

    Eix = xp: Eiy = yp: Eiz = zp

 BOTTOM = TPERP(1) * Eix + TPERP(2) * Eiy + TPERP(3) * Eiz

  IF BOTTOM = 0 THEN BOTTOM = .000001

    TOPSQ = (TPERP(2) * Eiz - TPERP(3) * Eiy) ^ 2
    TOPSQ = TOPSQ + (TPERP(3) * Eix - TPERP(1) * Eiz) ^ 2
    TOPSQ = TOPSQ + (TPERP(1) * Eiy - TPERP(2) * Eix) ^ 2
    TOP = SQR(TOPSQ)

    PHI = ATN(TOP / BOTTOM) * SGN(TPERP(2))

    CPHI = COS(PHI)
    SPHI = SIN(PHI)

PHS = (2 * PI / LAM) * XAROT

       APF = (ARP / RPDIS) ^ 2

IF withradome$ = "NO" THEN GOTO 2300

'- - - - Resolve Eaz and Eel into para. and perp.components

'- - - - Parallel polarization: POL = 1
    POL = 1
    GOSUB 4000

    REPARA = (EAZ * SPHI - EEL * CPHI) * COS(-PHS - IPD) * TRAN
    IMPARA = (EAZ * SPHI - EEL * CPHI) * SIN(-PHS - IPD) * TRAN
```

```
    XREPARA = (EAZ * SPHI + EEL * CPHI) * COS(-PHS - IPD) * TRAN
    XIMPARA = (EAZ * SPHI + EEL * CPHI) * SIN(-PHS - IPD) * TRAN

'- - - - Perpendicular polarization: POL = 0
    POL = 0
    GOSUB 4000

    REPERP = (EAZ * CPHI + EEL * SPHI) * COS(-PHS - IPD) * TRAN
    IMPERP = (EAZ * CPHI + EEL * SPHI) * SIN(-PHS - IPD) * TRAN
     XREPERP = (-EAZ * CPHI + EEL * SPHI) * COS(-PHS - IPD) * TRAN
    XIMPERP = (-EAZ * CPHI + EEL * SPHI) * SIN(-PHS - IPD) * TRAN

'- - - - CONVERT COMPONENTS BACK INTO Eaz AND Eel COORDINATES

        REAZ = REPARA * SPHI + REPERP * CPHI
        IMAZ = IMPARA * SPHI + IMPERP * CPHI
        REEL = -REPARA * CPHI + REPERP * SPHI
        IMEL = -IMPARA * CPHI + IMPERP * SPHI

        RAZP(JY, JZ) = RAZP(JY, JZ) + REAZ * APF
        IAZP(JY, JZ) = IAZP(JY, JZ) + IMAZ * APF
        RELP(JY, JZ) = RELP(JY, JZ) + REEL * APF
        IELP(JY, JZ) = IELP(JY, JZ) + IMEL * APF

        XREAZ = XREPARA * SPHI + -XREPERP * CPHI
        XIMAZ = XIMPARA * SPHI + -XIMPERP * CPHI
        XREEL = XREPARA * CPHI + XREPERP * SPHI
        XIMEL = XIMPARA * CPHI + XIMPERP * SPHI

        XRAZP(JY, JZ) = XRAZP(JY, JZ) + XREAZ * APF
        XIAZP(JY, JZ) = XIAZP(JY, JZ) + XIMAZ * APF
        XRELP(JY, JZ) = XRELP(JY, JZ) + XREEL * APF
        XIELP(JY, JZ) = XIELP(JY, JZ) + XIMEL * APF

    IF polarization$ = "H" THEN
        XRAZP(JY, JZ) = 0: XIAZP(JY, JZ) = 0
                END IF

    IF polarization$ = "V" THEN
        XRELP(JY, JZ) = 0: XIELP(JY, JZ) = 0
                END IF

    IF withradome$ = "YES" THEN GOTO 2400

2300    CPHS = COS(-PHS): SPHS = SIN(-PHS)

        REAZ = EAZ * CPHS
        IMAZ = EAZ * SPHS
        REEL = EEL * CPHS
        IMEL = EEL * SPHS

        RAZP(JY, JZ) = RAZP(JY, JZ) + REAZ * APF
```

```
        IAZP(JY, JZ) = IAZP(JY, JZ) + IMAZ * APF
        RELP(JY, JZ) = RELP(JY, JZ) + REEL * APF
        IELP(JY, JZ) = IELP(JY, JZ) + IMEL * APF

2400 RETURN

4000 '- - - - WALL PROGRAM SUBROUTINE (Planar Approximation) - - - -

THETA = ABS(THETA)
IF THETA > PI / 2 THEN THETA = THETA - PI

     CTH = COS(THETA)
     STH = SIN(THETA) * SIN(THETA)
     z(0) = 1

' - - - - - - - - - - - - - - - - - - - - - - - - - - - - - - - - - - - - - -

     FOR I = 1 TO (N.lay + 1)

       IF I <> (N.lay + 1) THEN 2980
       z(I) = 1
       GOTO 3040

2980   U3(I) = SQR(er(I) - STH)
       z(I) = CTH / U3(I)
       U3(I) = U3(I) * K0

3010   U2(I) = thk(I) * K0 * K0 * er(I) * ltan(I) / (2 * U3(I))

'      POL=0 IS PERPENDICULAR, POL=1 IS PARALLEL POL.

       IF POL = 1 THEN z(I) = 1 / (er(I) * z(I))

       R(1) = (z(1) - 1) / (z(1) + 1)

          IF I = 1 THEN GOTO 3050

3040      R(I) = (z(I) - z(I - 1)) / (z(I) + z(I - 1))

3050   NEXT I

     AR(1) = EXP(U2(1))
     AR(4) = 1 / AR(1)
     AR(2) = R(1) * AR(4)
     AR(3) = R(1) * AR(1)
     AI(1) = U3(1) * thk(1)
     AI(2) = -AI(1)
     AI(3) = AI(1)
     AI(4) = AI(2)

     FOR J = 2 TO N.lay
```

```
        BR(1) = EXP(U2(J))
        BR(4) = 1 / BR(1)
        BR(2) = R(J) * BR(4)
        BR(3) = R(J) * BR(1)
        BI(1) = U3(J) * thk(J)
        BI(2) = -BI(1)
        BI(3) = BI(1)
        BI(4) = BI(2)
        GOSUB 3430

      NEXT J

      BR(1) = 1
      BR(4) = 1
      BR(2) = R(N.lay + 1)
      BR(3) = BR(2)
      BI(1) = 0
      BI(2) = 0
      BI(3) = 0
      BI(4) = 0

      GOSUB 3430

      TRAN = 1

      FOR I1 = 1 TO N.lay + 1
        TRAN = TRAN * (1 - R(I1))
      NEXT I1
      TRAN = TRAN / AR(1)
      SUM = 0

FOR I1 = 1 TO N.lay
        SUM = SUM + thk(I1)
      NEXT I1
      TEST1 = AI(1)
      TEST2 = CTH * K0 * SUM

3393 IF TEST1 < TEST2 THEN TEST1 = TEST1 + 2 * PI

      IF TEST1 < TEST2 THEN GOTO 3393
      IPD = TEST1 - TEST2

RETURN

'- - - - - - - - - - - - - - - SUBROUTINE CMULT - - - - - - - - - - - - - - - - -
3430 FOR NA = 1 TO 3 STEP 2
        FOR NB = 1 TO 2

          R1 = AR(NA) * BR(NB)
          R2 = AR(NA + 1) * BR(NB + 2)
          E1 = AI(NA) + BI(NB)
          E2 = AI(NA + 1) + BI(NB + 2)
```

```
        R3 = R1 * COS(E1) + R2 * COS(E2)
        E3 = R1 * SIN(E1) + R2 * SIN(E2)
        CR(NB + NA - 1) = SQR(R3 * R3 + E3 * E3)
        IF (R3 > 0) AND (E3 > 0) THEN PA = 0
        IF (R3 < 0) THEN PA = PI
        IF (R3 > 0) AND (E3 < 0) THEN PA = 2 * PI
        Ci(NB + NA - 1) = (ATN(E3 / (R3 + 1E-12))) + PA

      NEXT NB
      NEXT NA

    FOR NA = 1 TO 4

      AR(NA) = CR(NA)
      AI(NA) = Ci(NA)

    NEXT NA

RETURN
```

Appendix E

Operations Manual for Tangent Ogive Radome Software

INSTALLATION

TANGENT OGIVE RADOME comes on a diskette containing the flat panel transmission module (WALL.EXE) and the tangent ogive radome module (RADOME.EXE). Both codes are in compiled format and do not require the user to have any type of interpreter or compiler on their machine to run.

Before installing and using the code, it is recommended that you make a backup copy of the diskette and keep it in a safe place.

When the INSTALL program is started, it will prompt the user for the name of the drive to which to install. If the drive is B: , it will be construed as another floppy disk. You can use this to make a backup copy. If it is C: or D: or E: , it will be automatically taken as a hard disk. When the program is installed to a hard disk, the program will automatically establish a "TORADOME" subdirectory into which it will load all operating modules.

When you install TANGENT OGIVE RADOME to a hard disk, you do not have to load the program from the diskette each time you wish to use it.

To install TANGENT OGIVE RADOME on a hard disk you must be in DOS (or DOS prompt if you are in Windows 95).

1. Make sure the TANGENT OGIVE RADOME program diskette is write-protected.
2. Insert the TANGENT OGIVE RADOME diskette in drive A; which should be a high-density (1.44 MB) disk drive.
3. Type **INSTALL** and press enter.
4. The program will then prompt you for a drive to which to install the program, as discussed above.

System Requirements For Version 1.0

All TANGENT OGIVE RADOME modules are designed to run on IBM compatible 386, or higher-level computers operating with DOS release 3.0 or later operating system. The program will also run with Windows 3.XX or Windows 95. The program is supplied on a 3.5-inch diskette, and requires a minimum of 4 MB of user memory (RAM).

To run TANGENT OGIVE RADOME, a VGA or SGVA high resolution color monitor is required.

Output to printers using a parallel interface can be directed to LPT1, LPT2, or LPT3. Output to printers using a serial interface can be directed to COM1 or COM2.

The TANGENT OGIVE RADOME program uses a math coprocessor if available. If a math coprocessor is not present, software math routines are used instead. However, computational run time is greatly extended.

Input/Output Requirements

A 3.5-inch hard disk, 1.44-MB drive and several blank formatted double-sided, double-density diskettes for saving data files.

Supported Printers

Most printers that support extended ASCII character sets such as Epson printers, and the like.

E.1 EXECUTING THE PROGRAM

To initiate use of the program, the user must enter the TORADOME subdirectory. This is accomplished by typing "TORADOME," and then ENTER.

The following modules are resident in the TORADOME subdirectory:

WALL.EXE Computes transmission through a multilayer dielectric wall.
RADOME.EXE Computes the boresight errors (BSE) and the copolarized and crosspolarized transmission loss of a radome enclosed monopulse antenna.

Description of these programs modules operations appear in subsequent subsections of this appendix.

E.2 FLAT PANEL ANALYSIS (WALL.EXE)

E.2.1 Theory

The flat panel transmission module (WALL.EXE) evaluates transmission coefficients (amplitude and phase) through a multilayer dielectric panel. The output data can be seen as a function of frequency, where angle-of-incidence is taken as a parameter. Angles-of-incidence of 0, 15, 30, 45, 60, 75, and 89 degrees are considered in the calculations. Data, are also computed for both the parallel and perpendicular polarization case.

The mathematics is developed as boundary value problem of forward and reverse propagating waves. From this formulation, a matrix equation results, based on the Fresnel reflections and transmissions at each boundary, which is evaluated using matrix multiplication.

E.2.2 Operation

From the TORADOME subdirectory, type "WALL.EXE," and then ENTER. The program will immediately come up on the screen. It is highly user interactive and easy to use with no study when the user provides the input data questions that come up on the screen. The program will prompt the user for the following information:

1. Computation low frequency (GHz);
2. Computation high frequency (GHz);
3. Computation frequency increment (GHz);
4. Number of layers, number of layers (dimensionless);
5. Thickness of the *I*th layer in inches, thk(I) (inches);
6. Relative dielectric constant of the *I*th layer, er(I) (dimensionless);
7. Loss tangent of the *I*th layer, ltan (I) (dimensionless).

The program then gives the user the choice of sending data to an output printer or just monitoring the results on their computer's display. A yes (y) or no (n) response is required.

E.3 MONOPULSE ANTENNA ANALYSIS

E.3.1 Theory

When a dielectric radome is placed over a monopulse antenna, the characteristics of the antenna are altered. The radome analysis module (RADOME.EXE) is an electromagnetic solution to this problem, and allows the user to assess the impact of (multilayer) radome wall design and radome shape on the performance of an enclosed monopulse antenna.

RADOME.EXE is a receiving-ray trace compilation. Tracing rays through the radome walls, the amplitude and phase associated with each ray is modified by the complex transmission coefficient of the radome wall for the ray. The ray bundles are then mathematically processed according to a radome analysis algorithm. In this program, the radome wall is assumed locally plane at the intercept point with parallel inner and outer walls. This simplification is known to produce reasonably accurate results.

The ray tracing module works for radomes with enclosed monopulse antennas of at least five wavelengths in diameter. RADOME.EXE also utilizes a parametric TANGENT OGIVE shape that allows the user to select a wide range of fineness ratios ranging from a hemisphere (L/D=0.5), to very high fineness ratios (L/D large).

The data files can be imported from most test editing graphing and worksheet programs, where the data can be edited, graphed, or incorporated into other applications.

E.3.2 Operation

From the TORADOME subdirectory, the user should type RAD and then press the ENTER key. Program RAD.EXE will automatically download.

The program is highly user interactive and easy to use with no study when the user provides the input data questions that come up on the screen. The program will prompt the user for the following information:

1. Computation frequency (GHz);
2. Number of radome layers, number of layers, (dimensionless);
3. Thickness of the *I*th layer in inches, thk(I), (inches);
4. Relative dielectric constant of the *I*th layer, er(I) (dimensionless);
5. Loss tangent of the *I*th layer, ltan (I) (dimensionless);
6. Antenna diameter (inches);
7. x coordinate of gimbal location (inches);
8. y coordinate of gimbal location (inches);
9. z coordinate of gimbal location (inches);
10. The delta-a (Δ_a) gimbal offset (inches). (For this parameter, please see the figure at the end of this appendix.);
11. The delta-b (Δ_b) gimbal offset (inches). (For this parameter, please see the figure at the end of this appendix.);
12. The user is then given the option of selecting an azimuth over elevation (AZ/EL) or elevation over azimuth (EL/AZ) gimbal configuration. Calculations are made for principal plane antenna scan angles only, beginning at on-axis (AZ=0, EL=0). The program will prompt the user for the following information:
13. The maximum gimbal scan computation angle (degrees). This will be the same for both azimuth and elevation scan;
14. The gimbal angle computational steps (degrees). This will also be taken as the same for both the azimuth and the elevation planes;
15. The radome length (inches);
16. The radome base diameter (inches).

The program then gives the user the choice of sending data to an output printer or just monitoring the results on his or her computer's display. A yes (y) or no (n) response is required.

After queries for inputs of all data parameters, RADOME.EXE proceeds to calculate data over the user specified angular ranges, for both vertical and horizontal antenna polarizations.

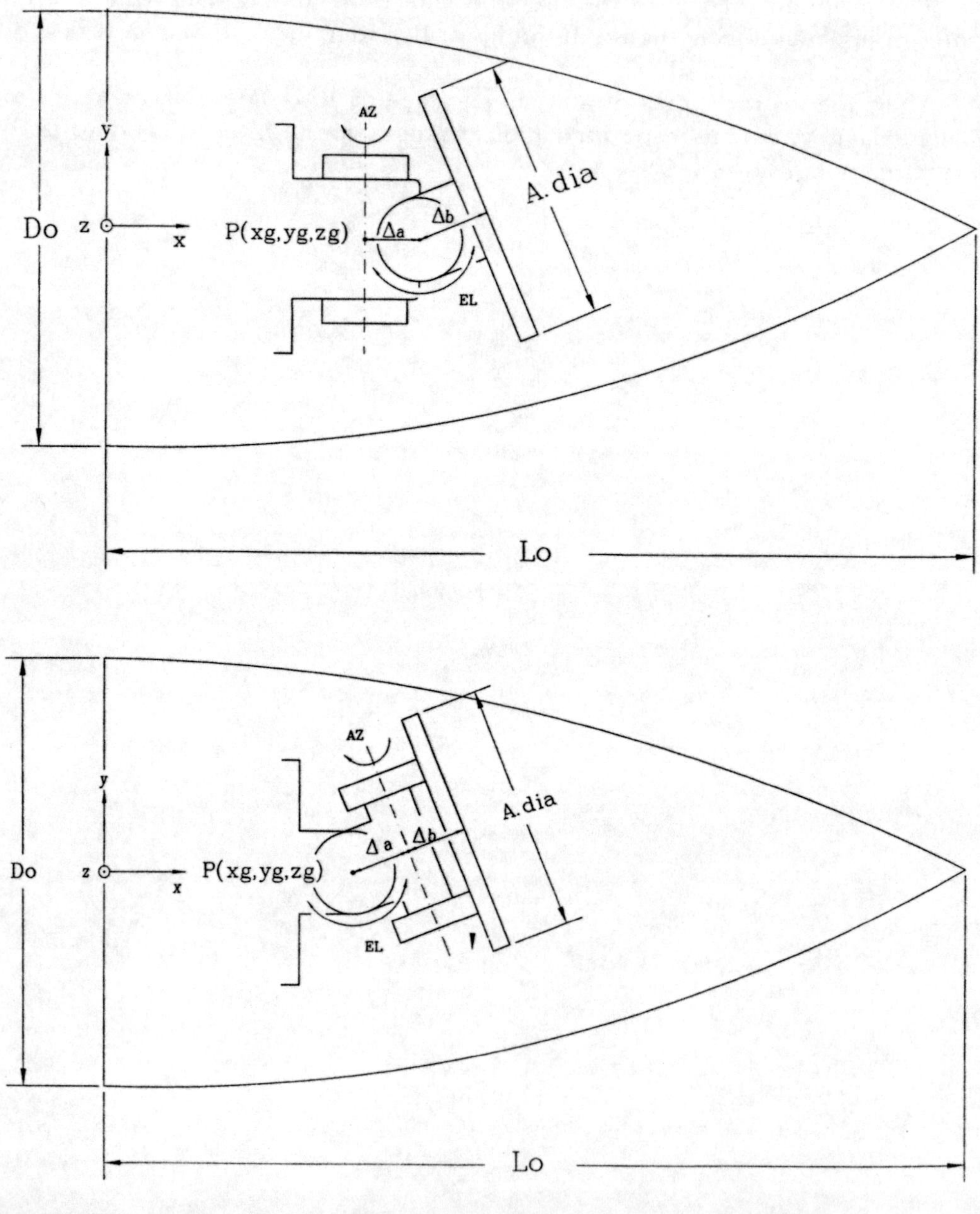

Figure E.1 Description of parameters used in RAD.

Glossary

Anisotropic A material whose permittivity and/or permeability are dependent on the direction of the applied fields. For instance, many useful crystals are anisotropic with applied magnetic fields.

Antenna bandwidth The range of frequencies within which the performance of the antenna, with respect to some characteristics, conforms to a specified standard.

Antenna beam The major lobe of the radiation pattern of an antenna.

Antenna boresight The tracking axis as determined by an electrical indication, such as the null direction of a conical scanning or monopulse antenna system, or the beam-maximum direction of a highly directive antenna.

Antenna directivity The value of the directive gain in the direction of its maximum value.

Antenna effective area In a given direction, the ratio of the power available at the terminals of a receiving antenna to the power per unit area of a plane wave incident on the antenna from that direction that is polarized coincident with the polarization that the antenna would radiate.

Antenna efficiency For an antenna with a specified planar aperture, the ratio of the maximum effective area of the antenna to the physical aperture area.

Antenna gain In a given direction 4π times the ratio of the radiation intensity in that direction to the net power accepted by the antenna from the connected transmitter.

Antenna noise temperature The temperature of a resistor having an available thermal noise power per unit bandwidth equal to that at the antenna output at a specified frequency. Noise temperature of an antenna depends on its coupling to all noise sources in its environment as well as noise generated within the antenna.

Antenna polarization In a given direction the polarization of the wave radiated by the antenna.

Antenna scan angle The angle between the maximum of the major lobe and a reference direction. In the case of missiles, the reference boresight is often chosen as the missile axis.

Antenna scanning A repetitive motion given to the major lobe of an antenna.

Aperture blockage A blocking of interference with the radiation from an aperture by obstacles such a metallic rain erosion tip.

Aperture illumination The field over the aperture as described by amplitude, phase, and polarization distributions.

Aperture of an antenna A surface, near or on an antenna, on which it is convenient to make assumptions regarding the field values for the purpose of computing fields at external points. The aperture is often taken as that portion of a plane surface near the antenna, perpendicular to the direction of maximum radiation, through which the major part of the radiation passes.

Array antenna An antenna comprising a number of radiating elements, generally similar, which are arranged and excited to obtain directional radiation patterns.

Axial ratio The ratio of the axes of the polarization ellipse.

Boresight error The angular deviation of the electrical boresight of an antenna from its physical boresight.

Conical scan A form of sequential lobbing in which the direction of maximum radiation generates a cone whose vertex angle is of the order of the antenna half-power beamwidth. Such scanning may be either rotating or notating according to whether the direction of polarization rotates or remains.

Cross-polarization The polarization that is orthogonal to the wave polarization that would be radiated by the antenna in a given direction.

Far-field region That region of the field of an antenna where the angular field distribution is essentially independent of the distance of the antenna. If the antenna has a maximum overall dimension D that is large compared to the wavelength, the far-field region is commonly taken to exist at distances greater than $2D^2/\lambda$ from the antenna, where λ is the wavelength.

Frequency band Institute of Electrical and Electronics Engineers (IEEE) radar frequency band designations:

L-band	1–2 GHz
C-band	4–8 GHz
X-band	8–12 GHz
K_u-band	12–18 GHz
K-band	18–27 GHz
K_a-band	27–40 GHz

Half-power beamwidth In a plane containing the direction of the maximum of a beam, the angle between the two directions in which the radiation intensity is one half (−3 dB) the maximum value of the beam.

Homogeneous A material whose permittivity and permeability does not depend on position.

Inhomogeneous A material whose permittivity and permeability does depend on position.

In-plane and cross-plane errors A monopulse guidance system processes the received antenna signals into azimuth and elevation angle error voltages that are proportional to the displacement of a target from the antenna axis. The missile autopilot uses these voltages to drive the control surfaces and steer the missile toward the target according to the guidance algorithm selected. When the target is exactly on the axis, these voltages are both zero. The effects of a radome are to contribute a BSE voltage component to each of the respective angle error channels.

If we move a gimballed seeker antenna along a principal plane, the BSE voltage component introduced to the corresponding scan plane angle error channel is called an *in-plane error*. The BSE error voltage contributor to the other angle error channel is called a *cross-plane error*. For instance, if we scan the antenna along the azimuth axis, the BSE voltage appearing in the azimuth angle error channel is an in-plane error. That appearing in the elevation angle error channel due to an azimuth-only motion of the antenna is a cross-plane error.

Isotropic A material whose permittivity and permeability are not dependent on the direction of the applied fields.

Lens antenna An antenna consisting of an electromagnetic lens and a radiating feed.

Linear array antenna An array antenna having the centers of the radiating elements lying along a straight line.

Look angle dependent The gimballed seeker antenna is generally controllable over some range of azimuth and elevation angles (AZ, EL). We say the BSE is look angle dependent if it is not a constant but, in fact, a function of the azimuth and elevation angle, that is, BSE(AZ, EL).

Mainlobe The radiation lobe containing the directions of maximum radiation.

Magnetron A diode vacuum tube in which the flow of electronics is controlled by an externally applied magnetic field to generate power at microwave frequencies. These tubes first made it possible to design and develop high-power, long-range, pulsed radars.

Monopulse A form of simultaneous lobbing whereby direction-finding information is obtainable from a single pulse.

Paraboloidal reflector A reflector that is a portion of a paraboloid of revolution.

Phase center In a given direction and for a specified polarization, the center of curvature of the wavefront of the radiation from an antenna in a given plane.

Phased antenna array An array antenna whose beam direction or radiation pattern is controlled primarily by the relative phase of the excitation coefficients of the radiating elements.

Planar antenna array An array antenna having the centers of the radiating elements lying in a plane.

Proportional guidance trajectory A proportional missile guidance algorithm constrains the missile to fly as near as possible along a line that has a fixed angle relative to the line-of-sight (LOS). For constant missile and target velocity and without a target maneuver, with proportional guidance the missile is flying a fixed course toward the predicted impact point and the proportional guidance trajectory is a straight line. With a target maneuver, the guidance system drives the missile control surfaces to drive the rate-of-change of this angle to zero. The

G's the missile must pull toward the end of the engagement does not substantially increase as it does for a pursuit guidance system.

PT boat Patrol torpedo boat. This is a type of high-speed navy attack boat used in World War II. One of the primary weapons carried by this craft was the torpedo.

Pursuit guidance trajectory A pursuit missile guidance algorithm constrains the missile to fly as near as possible to the line joining the missile seeker and the target. We call this line the line-of-sight (LOS). The guidance system and missile autopilot maneuver the missile to drive the angle between the LOS and the missile flight axis to zero. When the missile is flying toward a moving target, the pursuit guidance trajectory is a curved one. The curvatures, and therefore the G's the missile must pull, become very high toward the end of the engagement.

Radiating element A basic subdivision of an antenna which in itself is capable of effectively radiating or receiving radio waves. Typical examples of a radiating element are a slot or dipole antenna.

Radiation efficiency The ratio of the total power radiated by an antenna to the net power accepted by the antenna from the connected transmitter.

Radiation pattern A graphical representation of the radiation properties of the antenna as a function of space coordinates.

Sidelobe A radiation lobe in any direction other than that of the mainlobe.

Sidelobe level The level of the highest sidelobe relative to the main beam.

Spiral antenna An antenna consisting of one or more conducting wires arranged as a spiral. Spiral antennas are usually classified according to the surface to which they conform, that is, conical spiral or planar spiral, and according to mathematical form such as equiangular and Archimedean.

Thin-wall radome Single-wall (monolithic) radomes whose overall electrical thickness is equal to, or less than, a tenth of a wavelength.

Tracking A motion given to the major lobe of an antenna such that a selected moving target is contained within the major lobe.

Vertically polarized wave A plane wave in which the electric vector is in the vertical plane containing the direction of propagation.

List of Symbols and Abbreviations

A	=	Antenna aperture physical area (m^2)
A_{eff}	=	Antenna aperture effective area (m^2) = $\eta_a A$
ARH	=	Antiradiation homing
AZ	=	Azimuth angle (radians)
A_0–A_5	=	Constants used in fifth-order polynomial regression (dimensionless)
B	=	Magnetic induction (Wb/m^2)
B_0	=	Parameter used to describe tangent ogive shape (m)
B_L	=	Space frame radome beam length (m)
B_s	=	Systems noise bandwidth (Hz)
B_i, C_i	=	Reverse and forward propagating waves, respectively, in the i'th layer of a multilayer dielectric (V/m)
BSE	=	Boresight error (mrad)
BSES	=	Boresight error slope (mrad/radian)
c	=	Speed of light in vacuum = 3×10^8 m/s
C_r	=	Concentration of rain (g/cm^3)
°C	=	Degrees centigrade
C_i	=	Convergence constant (dimensionless)
C_D	=	Drag coefficient (dimensionless)
d	=	Sample-to-sample spacing or infinitesimal current length (m)
d_x, d_y	=	Sample spacing in the x and y planes, respectively (m)
D	=	Electric displacement (C/m^2)
D_a	=	Antenna diameter (m)
D_r	=	Rain drop diameter (m)
$\overline{D}$	=	A vector in the surface plane of a radome that is orthogonal to the plane formed by the surface normal and the propagation vector (m)
DF	=	Divergence factor
D_o, D_i	=	Radome outer and inner base diameters, respectively (m)
E	=	Electric field intensity (V/m)

E_θ, E_ϕ = Components of E in the θ and ϕ directions, respectively (V/m)
E^i = Incident plane wave function (dimensionless)
E_x^i, E_y^i, E_z^i = Components of incident plane wave function in the x, y, and z directions, respectively (dimensionless)
E^{AF} = Array factor (dimensionless)
$E_\theta^{AF}, E_\phi^{AF}$ = Components of E^{AF} in the θ and ϕ directions, respectively (dimensionless)
ECM = Electronics countermeasures
ERF = Error function (optimization constant)
E_t = Transmit mode array factor pattern (dimensionless)
EM = Electromagnetic
EL = Elevation angle (radians)
f = Frequency (Hz)
F = System noise figure (dB)
F^a = Complex valued aperture distribution on the antenna surface (dimensionless)
F_{mn}^a = Value of F^a at the mn'th point on the antenna surface (dimensionless)
$F_{mn}^{a'}$ = Value of aperture distribution on the mn'th point of an external antenna surface or plane (dimensionless)
F_o, F_i = Fineness ratio based on outer and inner radome dimensions, respectively (dimensionless)
FFT = Fast Fourier transform
G = Antenna gain (dimensionless) = $\eta_a G_D$
$\overline{G}$ = A vector that is orthogonal to the plane formed by $\overline{D}$ and the propagation vector (m)
G_D = Antenna directivity (dimensionless)
G_s = System power gain (dimensionless)
h_B = Space frame radome beam dimension in radial direction (m)
H = Magnetic field intensity (A/m)
H_θ, H_ϕ = Components of H in the θ and ϕ directions, respectively
HPSN = Hot pressed silicon nitride
$I(z)$ = Current distribution in z direction (A)
I_o = Antenna current maximum amplitude (A)
I_{mn} = Current at the mn'th sample point (A)
IR = Infrared
IPD = Insertion phase delay (radians)
$[I]$ = Identity matrix (dimensionless)
J = Integer used as an index in the convergence algorithm (dimensionless)
k = Wave number (m^{-1})
k_B = Boltzmann's constant = 1.38×10^{-23} W-s/K
K = Kelvin

$\overline{\mathbf{K}}$ = Propagation vector (m)
K_{az}, K_{el} = Azimuth and elevation monopulse channel sensitivity constants, r espectively (V/mrad)
K_m = Erosion constant for a particular radome material (dimensionless)
L = Radome power transmission loss factor (dimensionless)
L_B = Space frame radome frame blockage loss (dB)
L_{dB} = Radome transmission loss (dB) = 10 log L
LNA = Low noise amplifier
L_a = Linear antenna dimension (m)
L_o, L_i = Radome outer and inner lengths, respectively (m)
LSY = Lanthana strengthened yittria
m, n, p = Integers orders or indices (dimensionless)
m_s = Slope of radome at ray intercept point (dimensionless)
M, N = Total number of sample points in orthogonal coordinate directions, respectively (dimensionless)
MOM = Method of moments
M_N = Mach number (dimensionless)
$\overline{N}$ = Surface normal vector (m)
n_x, n_y, n_z = Scalar components of the surface normal vector in the x, y, and z directions, respectively (m)
NF = System noise factor (dimensionless)
p = Antenna distribution pedestal parameter (dimensionless)
$P(\theta,\phi)$ = Antenna power radiated per unit solid angle (W/rad^2)
PC = Personal computer
PWS = Plane wave spectra
P_{si}, P_{so} = Signal power in and out of system, respectively (W)
P_{ni}, P_{no} = Noise power in and out of system, respectively (W)
q = Antenna distribution power (dimensionless)
r_a = Antenna radius (m)
r_P = Radius of P'th current ring on spiral antenna aperture (m)
R_o = Parameter used to describe tangent ogive shape (m)
R_w = Radome front face voltage reflection coefficient (dimensionless)
R_{wi}, T_{wi} = Fresnel reflection and transmission coefficients, respectively, at the i'th interface of a multilayer dielectric wall (dimensionless)
RSSN = Reaction scintered silicon nitride
SCFS = Slip cast fused silica
T = Physical temperature (K)
T_a = Antenna noise temperature (K)
T_o = Ambient temperature = 290 K
T_e = System effective noise temperature (K)
t_i = Thickness of i'th layer (m)
T_k = Material eroded from radome surface (μm)
T_r = Physical temperature of radome (K)

T_w = Radome voltage transmission coefficient (dimensionless)
$\tan\sigma$ = Loss tangent of medium (dimensionless) = ϵ''/ϵ'
$\tan\sigma_i$ = Loss tangent of *i*'th layer (dimensionless)
V = Velocity (m/s)
V_a = Receive mode antenna terminal voltage (V)
V_{az}, V_{el} = Monopulse antenna azimuth plane and elevation plane error voltages, respectively (V)
V_R, V_F = Volume of resin and reinforcement fibers, respectively (m^3)
$\bar{x},\bar{y},\bar{z}$ = Cartesian coordinates unit vectors in the *x*, *y*, and *z* directions, respectively (m)
x_g, y_g, z_g = Antenna gimbal rotation center coordinates (m)
W_B = Space frame radome beam width (m)
Z_i = Characteristic wave impedance of the *i*'th layer of a multilayer dielectric wall (ohms)

Greek

α = Attenuation constant ($m^{\circ 1}$)
α_m = Erosion constant for a particular radome material (dimensionless)
α_r = Rain impingement angle relative to the radome surface normal (radians)
β = Phase constant (m^{-1})
γ = Propagation constant (m^{-1}) = $\alpha + j\beta$
γ_i = Propagation constant in the *i*'th layer of a multilayer dielectric wall (m^{-1})
σ = Differential path length (m)
Δ_{IPD} = Difference between the parallel and perpendicular insertion phase delays, respectively (rad)
Δ_{az}, Δ_{el} = Monopulse antenna azimuth plane and elevation plane difference channel signals, respectively (V)
ϵ = Permittivity of medium (F/m)
ϵ' = Real part of permittivity (F/m)
ϵ'' = Imaginary part of permittivity (F/m)
ϵ_F = Relative dielectric constant of reinforcement fibers (dimensionless)
ϵ_m = Relative dielectric constant of mixture (dimensionless)
ϵ_r = Relative permittivity of medium (dimensionless)
ϵ_{ri} = Relative permittivity of *i*'th layer (dimensionless)
ϵ_R = Relative dielectric constant of resin (dimensionless)
ϵ_0 = Permittivity of free space = 8.854×10^{-12} (F/m)
η = Wave impedance of medium (ohms) = $\sqrt{\mu/\epsilon}$
η_a = Antenna radiation efficiency (dimensionless)
$\bar{r},\bar{\theta},\bar{\phi}$ = Spherical coordinates unit vectors in the *r*, θ, and ϕ directions, respectively (m and radians)

θ_B = Brewster angle (radians)
Θ_o, Φ_o = Antenna elevation and azimuth plane half-power beamwidths, respectively (radians)
λ = Wavelength (m)
λ_0 = Wavelength in free space (m)
μ = Permeability of medium (H/m) = $\mu_r \mu_0$
μ_r = Relative permeability of medium (dimensionless)
μ_0 = Permeability of free space = $4\pi \times 10^{-7}$ (H/m)
$\perp, \|$ = Perpendicular and parallel, respectively (dimensionless)
ρ = Radius length parameter (m)
σ_x, σ_y, σ_z = Phase angle of the *x*, *y*, and *z* components of incident field (radians)
Σ = Monopulse antenna sum channel signal (V)
ψ = Cumulative angle around spiral antenna center (radians)
Ψ, Φ = Angle parameters that describe orientation of surface normal vector at ray intercept point (radians)
ω = Radian frequency (radians/s) = $2\pi f$

About the Author

Dennis J. Kozakoff has more than thirty years of experience related to antennas, radomes, and electromagnetics. He is the founder of Millimeter Wave Technology, Inc. (MWT) of Marietta, GA, a high-technology manufacturer of instrumentation and electromagnetic related materials. Previously, he was a program manager at the Electromagnetics Laboratory, Georgia Tech Research Institute, and a senior staff engineer at the Martin Marietta Corporation, Orlando, FL, and served in engineering roles at the Sperry Rand Corporation and North American Aviation.

He holds a B.S.E.E. from the Polytechnic Institute of Brooklyn, an M.S.E. degree from the University of Florida, and a Dr. Engineering degree from Southeastern Institute of Technology. He is a senior member of the Institute of Electrical and Electronic Engineers and a member of the IEEE Antennas and Propagation Society and, in the past, has served as vice chairman of the IEEE Professional Group on Microwave Theory and Techniques, Atlanta, GA (1983).

In addition, Kozakoff is a member of the International Society for Optical Engineering (SPIE). He served as session chairman at the SPIE LASE '94 Symposium in Los Angeles, CA, (1994) and at the International Conference on Infrared and Millimeter Waves in Orlando, FL, (1995), both of which concerned millimeter-wave electronics. He was general chairman of the 15th Symposium on Electromagnetic Windows, Georgia Institute of Technology, Atlanta, GA, and was an invited member on the Antenna Technology Panel, International Symposium on Solar Power Satellites, Toulouse, France (both in 1980). Kozakoff was a recipient of an NASA Award for Multipaction Research (1971) and is a holder of several U.S. patents related to antennas and radomes, a nominated member to the Phi Kappa Phi honorary society, and a member of the New York Academy of Sciences.

He has written or coauthored more than 150 technical publications, including "Seeker Antennas," Chapter 38 of the *Antenna Engineering Handbook*, 3rd Edition, edited by R.C. Johnson (New York, NY: McGraw Hill Book Company).

Index

Acrylic, 59
Additive system noise, 62–63
Aerodynamics drag, 120–21, 123
Aerothermal heating, 176–77
Air/dielectric interface, 8, 75
Air-supported radome, 192
Alumina, 53
Amplitude distribution, 201
Anechoic chamber, 185
Angle-of-arrival, 6
Angle-of-incidence, 69, 78, 104, 113, 121–24, 192
Anistropy, 114
Antenna modeling, 109–10
Antiradiation homing system, 41, 183, 185
Aperture current, 30–31
Aperture distribution
  defined, 33
  geometrical optics, 103–4
  linear, 32–37
Aperture integration, 173
Aperture point transformation, 170, 172
Aperture sample point, 172
Apparent permittivity, 210–11
Archimedean spiral antenna, 41–42, 184
ARH system. *See* Antiradiation homing system
Array factor, 31–33, 37, 190, 198–99, 201, 213, 218–19
A-sandwich radome
  bulkhead reflection, 113–14
  space-frame radome, 195
  wall material/design, 4, 49–53, 73, 78–81
Attenuation constant, 16
Autoclave molding, 48
Auxiliary relations, 15
Axial symmetry, radome, 134–39
Azimuth angle, 19–24, 170, 172

Backward ray tracing, 102–3, 111
BASIC, 167
Bayliss distribution, 201
Beamshape synthesis, 33
Beam steering, 200
Beryllia, 53
Blackbody radiation, 60
Blockage loss, 194–96
Boresight error, 6–7, 9, 73, 78–79, 97, 102, 113, 153, 169–70, 173, 176–77, 185–86, 196
  defined, 157
Boresight error slope, 6–7, 9, 73, 78–79, 153, 157, 173, 185–86
Boresight null, 158, 169, 189
Boron aluminate, 58
Boron nitride, 53
Borosilicate glass, 53
Brewster angle, 78
Broadband radome, 73
BRUTE, 120, 131, 140–42
B-sandwich radome, 73, 81
BSE. *See* Boresight error
BSES. *See* Boresight error slope
Bulkhead reflection, 111, 113

Calcium lanthanum sulfide, 58
Cartesian coordinate system, 11–14, 19–24, 28, 158, 161–62, 205–6
Cascaded system, 60–61
Cavity-backed spiral antenna, 184, 189–90
Celsian, 53–54
Ceramic material, 4, 53–54, 58–59
Circular aperture, 37–38, 161–62
Circular patch radiator, 160–61
Circular polarization, 7, 17, 42, 72, 138, 183
Classes, radome, 73
Closing velocity, 155
Column matrix, 18
COMPOSE, 120, 149–51
Computer-aided design, 97–98
Computer run time, 9
Conductivity term, 207
Conscan (conical scan) antenna, 177
Coordinate system, 11–14, 19–24, 28, 158, 161–62, 205–6
Copolarization, 97
Core material, 50–53
Corporate feed, 199
Cosine aperture distribution, 35
Cosine factor, 163
Coupling changes, 32
Cramer's rule, 18
Cross-polarization, 72, 97
C-sandwich radome, 50, 81–82
Curl, vector, 14, 207–8, 213, 215
Curl of curl, vector, 14
Current element, 30–31
Cylindrical coordinate, 205–6

Dacron, 192
Depolarization, 6–7, 9
DF. *See* Divergence factor
DF system. *See* Direction finding system
D-glass, 49–51
Dielectric constant, 47
Dielectric loss, 7–8, 16, 60–63, 193–96
Dielectric properties, fiber/resin, 49–53
Dielectric space frame radome, 196–98
Dielectric wall material
  dual-mode, 56–63
  inorganic, 53–56
  organic, 47–53
Diffraction theory, 189
Dipole-array antenna, 3
Dipole radiator, 31–32
Dirac delta functions, 217
Directional guidance radome, 73
Direction finding system, 41
Directivity
  fundamentals, 27–29
  linear aperture, 34, 36
Direct ray method. *See* Geometric optics
Distributive law, 13
Divergence factor, 14, 114, 213, 215
DOME, 193
Dual-mode radome wall, 56–56
  nonorganic, 57–58
  organic, 59
Dual-polarized wideband antenna, 5
Duroid, 176

ECM. *See* Electronics countermeasure
Effective antenna area, 29
E-glass, 49–51
Electric field linear aperture, 32
Electromagnetic field, 99, 102, 213
Electromagnetic sensor, 5–6
Electromagnetic theory, 14–17
Electromagnetic Window Symposium, 5
Electronics countermeasure, 183, 185
Elemental current radiation, 213–19
Element pattern factor, 32
Element radiator, 198
Elevation angle, 19–24, 170, 172
Elliptical aperture, 38
Elliptical polarization, 17
Ellipticity ratio, 97
EM field. *See* Electromagnetic field
Equiangular spiral antenna, 41
Equivalent noise temperature, 60
Equivalent transmission coefficient, 187
ERF. *See* Error function
Error function, 128–34, 136
Error voltages, monopulse antenna, 155, 157
  antenna without radome, 158–68
  antenna with radome, 169–77
Expansion coefficient, 41
External reference plane, 107–8, 187–88

Far-field radiation, 27–28, 30, 32, 104, 157, 169, 173
  GO transmitting mode, 104–5
  infinitesimal dipole, 213, 216–18
  line source, 34–37
  physical optics, 108

two-dimensional aperture, 37–40
Fast Fourier transform, 102, 107, 109
FFT. *See* Fast Fourier transform
Fiberglass, 78, 192
Fiber radome wall, 49–50
Fineness ratio, 120–21, 123, 153
Flat panel analysis, 240–41
Flat-plate antenna, 39–40, 158
Flat-plate voltage transmission coefficient, 136–37
Foam, 194–96
Four-arm spiral antenna, 43–44
Fourier transform, 33, 37, 104, 109
fast, 102, 107, 109
Frame blockage loss, 193–96
Free-space permittivity, 213
Free-space wave impedance, 31
Frequency independence, 41
Fresnel reflection, 70, 211
Front face voltage reflection, 71

Gain
fundamentals, 27–29
linear aperture, 36
Gallium arsenide, 57–58
Gauss's elimination method, 135
General Electric, 58
General theory of diffraction, 189
Geodesic dome, 193
Geometric optics, 98
analysis features, 100, 102
receiving mode, 102–4
transmitting mode, 104–5
versus physical optics, 98–100, 105, 107
Germanium mullite, 58
Gimbal-mounted antenna, 170–72
Gimbal-mounted seeker head, 155, 157
Gimbal relationships, 19–24
Glass, 52–53
GO. *See* Geometric optics
Grating lobe, 36–37
Green's function, 108
GTD. *See* General theory of diffraction
Gyros, 157

Hafnium titanate, 58
Half-power beamwidth, 27, 29, 35–36, 158, 183
Half-wave wall radome, 73, 75–78, 185
Hand rules, 42
Hardware-in-the-loop, 7, 185
Helmholtz wave equation, 216
Hemispherical radome, 120–21, 192
High-speed missile radome, 4
Honeycomb material, 50, 52, 78, 81
Hot pressed silicon nitride, 53–54
HPBW. *See* Half-power beamwidth
HPSN. *See* Hot pressed silicon nitride
Huygens's principle, 27, 98–99, 105, 107, 187
HWIL. *See* Hardware-in-the-loop
Hydrophobic coating, 63
Hypalon, 192
Hypersonic missile, 53

Identity matrix, 18
Incidence angle, 69, 78, 104, 113, 121–24, 192
Infinitesimal dipole, 162–63, 213–18
Infrared sensor, 56–57
Inorganic radome dielectric wall, 53–56
In-phase quadrature, 158
Insertion loss, 6–9, 194
Insertion phase delay, 71–72, 79
software program, 85–95
Instantaneous field of view, 183
Institute of Electrical and Electronic Engineers, 138
Interactive algorithm, 127
Intercept point, 97, 103–4, 114, 126
determination methods, 127–34
POLY location method, 136
wave decomposition, 136–39
Interferometer, 183
Internal reflection, 110–14
Inverse transformation, 18–19
IPD. *See* Insertion phase delay
IR sensor. *See* Infrared sensor
Iterative algorithm, 127–34

Kevlar, 49–51
Kirchoff-Fresnel integral, 105
Korex honeycomb, 50

Lanthana-strengthened yttria, 58
Layer wave impedance, 70
Least squares method, 134–35
Left-hand circular polarization, 7, 138
Lens antenna, 158
LHCP. *See* Left-hand circular polarization
Linear aperture, 32–37
Linear phase shift, 200
Linear polarization, 17
Line-of-sight, 155
Line source antenna, 33–38
LNA. *See* Low-noise amplifier

Log-spiral antenna, 41
Loss. *See* Dielectric loss; Frame blockage loss; Insertion loss; Ohmic loss; Transmission loss
Loss tangent, 16, 47, 50, 54–55, 57, 71, 176, 207, 210
Low-frequency radome, 73
Low-noise amplifier, 61
LSY. *See* Lanthana-strengthened yttria

Magic T, 201
Magnetic field linear aperture, 33
MAS. *See* Methyl methacrylate styrene
Matrix algebra, 17–19
Maxwell's equations, 14–16, 38, 207–10
Metal space-frame radome, 192–96
Method of moments, 100, 108, 189
Methyl methacrylate styrene, 59
Methylpentene, 59
Microstrip-array antenna, 158, 160–61
Microwave magnetron, 3
Microwave sensor, 56
Millimeter-wave sensor, 56
MIL-R-7705B radome, 72–73
Miss distance, 7
Missile guidance system
  antiradiation homing, 41
  boresight error, 7
  monopulse antenna, 154–57
  spiral antenna, 183
Missile-seeker antenna, 21
Modeling factors
  aerothermal heating, 176–77
  antenna design, 109–10
  conscan antenna, 177
  radome wall, 114
  rain erosion, 175–76
MOM. *See* Method of moments
Monolithic wall radome. *See* Half-wave wall radome
Monopulse antenna, 102
  block diagram, 159
  boresight error, 155–57
  features, 153–55
  functions, 155
  radome modeling, 173–77
  software program, 242–44
  steering, 155–57
  sum/difference patterns, 186, 201
  with radome, 169–73
  without radome, 158–68
Mullite, 58
Multilayer wall radome
  ARH/ECM antenna, 185
  C-sandwich, 81–82
  features, 79, 81–84
  seven-layer, 83–84
  style definitions, 73
  transmission equations, 69–72
Multimode spiral antenna, 188–90
Multiple-frequency band radome, 73

Narrowband radome, 73
Neoprene, 192
NF. *See* Noise factor
Nitroxyceram, 53–54
Noise factor, 60
Noise figure, 60
Noise temperature, 60–63
Nomex honeycomb, 50
Nomograph, 97
Normal incidence, 192, 208, 211
Nose radome, 3
Nylon, 59, 192
Nyquist criteria, 36

Office of Naval Research, 58
Ogive radome
  secant, 120
  tangent, 120–27
Ohmic loss, 29, 78
On-axis target, 123–24
One-dimensional aperture transform, 38
ONR. *See* Office of Naval Research
Open-loop testing, 185
Optical blockage function, 195–96
Optical material transmittance, 57
Organic radome dielectric wall, 4, 47–48
  core materials, 50–53
  sandwich type, 48–50
  solid, 48
Orthogonal polarization, 17, 138
Out-of-phase quadrature, 158

Parabola antenna, 190–91
Parallel polarization, 17, 71–72, 77–78, 80–82, 84, 110, 137, 211
Patch radiator, 158, 160–61
PATTERN, 165–67, 179–81
Pattern look direction, 172
PC. *See* Personal computer
Pedestal parameter, 35–36
Performance, radome, 8–10

Permeability, 213
Permittivity, 15–16, 55, 71, 176, 210–11, 213
Perpendicular polarization, 17, 70–72, 76, 79–80, 82–83, 110, 137, 211
Personal computer, 10, 98
Phase-array antenna, 198–201
Phase constant, 16
Phase quadrature, 158
Phenolic honeycomb, 78, 81
Physical optics, 98–100
 receiving mode, 100, 102–7, 187–88
 transmitting mode, 106
Planar aperture, 37–40
Planar radome, 190
Planar spiral antenna, 41–43
Plane wave, Huygens's principle, 98–99
Plane wave spectra, 100, 108–9
Plastics, 48, 50, 59, 78, 196
Plywood radome, 3–4
PO. *See* Physical optics
Pointing error, 193, 196
Point source, 98–99
Polarization
 array factor, 31–32
 circular, 7, 17, 42, 72, 138, 183
 copolarization, 97
 cross-polarization, 72, 97
 electromagnetic theory, 17
 elliptical, 17
 linear, 17
 orthogonal, 17, 138
 parallel, 17, 71–72, 77–78, 80–82, 84, 110, 137, 211
 perpendicular, 17, 70–72, 76, 79–80, 82–83, 110, 137, 211
Pollucite, 58
POLY, 120, 135–36, 145–49
Polycarbonate, 59
Polyethylene, 49–51
Polymethyl methacrylate, 59
Polystyrene, 59
Polyurethane foam, 52–53
Power series radome, 120–21
Prediction accuracy, 110–14
Pre-preg fabrication, 49
Printed circuit technology, 183
Propagation, 97, 209–10
PWS. *See* Plane wave spectra
Pyroceram, 53–55

Quartz, 49–51, 54, 81

Radiation, current element, 30–31, 213–19
Radiation efficiency, 29
Radiation intensity function, 27–28
Radio frequency, 3, 56
Radome, defined, 3
RADOME software, 171, 173–74, 239–44
Rain erosion, 63, 81, 108, 175–76
Rain erosion cap, 176
Range-to-target, 155
Ray bounce error, 113
Rayceram, 53–55
Ray tracing, 97, 99, 105–7
 backward, 102–3, 111
 computational error, 109–14
 intercept point, 127–34
 modeling issues, 119–20
 radome with axial symmetry, 134–39
 spiral antenna, 186–88
 tangent ogive radome, 120–27
Ray vector transformation, 172
Reaction sintered silicon nitride, 53–54
Receiving mode
 geometric optics, 102–4
 physical optics, 100, 105–7
Rectangular aperture, 38
Rectangular coordinates, 205
Reference spherical coordinate system, 22
Reflector antenna, 158
Refractive index, 58
Relative dielectric constant, 47
Resin, 49–52, 54
RF. *See* Radio frequency
RHCP. *See* Right-hand circular polarization
Right-hand circular polarization, 7, 42, 138
Rootfinder algorithm, 129–34
ROOTS, 120, 132, 142–45
RSSN. *See* Reaction sintered silicon nitride

Sample spacing, 37, 127, 161
SAN. *See* Styrene acrylonitrile
Sandwich wall radome, 4, 6, 48–50
 B-type, 73, 81
 core materials, 50–53
 C-type, 50, 81–82
 defined, 69
 See also A-sandwich radome
SATCOM antenna. *See* Satellite communications antenna
Satellite communications antenna, 3–4, 7, 34, 63
Scalar function, 215

Scalar multiplication, 13
Scalar triple product, 13
Scattered energy, 111–12
SCFS. *See* Slip cast fused silica
Secant ogive radome, 120
Seeker antenna. *See* Monopulse antenna
Servo loop, 157
S-glass, 49–50
Shape, radome wall
  choice rationale, 120–23
  mathematical modeling, 123–25
Sidelobe antenna, 6
Sidelobe level, 6–7, 9, 33–36, 97, 201
Sidewall reflection, 110–14
Silicon nitride, 54, 58, 175
Sinusoidal time equation, 209
Slip cast fused silica, 53–55, 175
Snell's law, 114
Software, multilayer dielectric wall, 85–95
Solid radome wall, 48
Space frame radome, 192–93
  dielectric material, 196–98
  dielectric wall loss, 194–95
  frame blockage loss, 195–96
  pointing error, 196
Space phase term, 170, 218
Spectra, 49–51
Sperry Rand Corporation, 3
Spherical coordinate system, 28, 158, 161–62, 206
Spiral antenna, 41–43
  multimode, 188–90
  technology, 183–88
Square aperture, 38
Square patch radiator, 158, 160
Steering, missile, 155, 157
Styles, radome, 73
Styrene/styrene acrylonitrile, 59
Surface integration method, 100, 102–7, 187–88
Surface normal vector, 125–27, 136

Tangent ogive radome, 120–27
  intercept point, 127–34
  program listing, 221–37
  shape characteristics, 120–23
  shape modeling, 123–25
  software manual, 239–44
  surface normal vector, 125–27
Tapered aperture, 35
Teflon, 63, 176, 192
Telecommunications antenna, 190–91
Television receive-only, 61
TEM wave. *See* Transverse electromagnetic wave
Texas Instruments, 58
Thermal analysis, 176–77
Thin-wall radome, 73, 78–80
Three-dimensional geometric optics, 99, 102, 108–9
Time-domain equation, 207
Time harmonic field, 213
Tip modeling, 109–10
Tracking loop, 155
Transmission, radome wall
  efficiency, 72
  equations, 69–72
  model/computational error, 110
  software program, 85–95
Transmission coefficients, radome
  circular polarization, 72
  equations, 211
  flat-plate antenna, 136–37
  goemetrical optics, 104
  physical optics, 105, 107
  spherical radome, 114
  spiral antenna, 186–188
  voltage, 71, 75–78, 110, 172–173, 201
Transmission loss, 97, 173, 193–95, 198
Transmitting mode
  geometric optics, 104–5
  physical optics, 106–8
Transverse electromagnetic wave, 208
TVRO. *See* Television receive-only
Two-dimensional aperture, 37–40
Two-dimensional geometric optics, 99, 102
Two-dimensional method of moments, 108
Types, radome, 73

UAV. *See* Unmanned aerial vehicle
Unity factor, 29, 35, 163
Unmanned aerial vehicle, 63
URLIM, 176

Vacuum bag fabrication, 48–49
Vector curl, 14, 207–8, 213, 215
Vector field, 137–39
Vector intercept, 129–30
Vector mathematics, 11–14, 164
Vector multiplication, 13
Vector triple product, 13
Voltage reflection coefficient, 71
Voltage standing wave ratio, 97, 190

Voltage transmission coefficients, 71, 75–78, 110, 136–137, 172–173, 201
Von Karman radome (query van), 120–21, 127
VSWR. *See* Voltage standing wave ratio

WALL, 173, 192, 195, 239–41
Walls. *See* Dielectric wall material; Transmission, radome wall
Wall variation, radome, 109–10
Wave components, radome, 207–10
Wave decomposition, 136–39
Wave impedance, 16, 31, 70, 211
Wet layup, 49

Yagi antenna, 3
Yttria, 58

Zinc germanate, 58
Zinc selenide, 57–58
Zinc sulfide, 57–58

The Artech House Antenna Library

Helmut E. Schrank, *Series Editor*

Advanced Technology in Satellite Communication Antennas: Electrical and Mechanical Design, Takashi Kitsuregawa

Analysis Methods for Electromagnetic Wave Problems, Volume 2, Eikichi Yamashita, ed*itor*

Analysis of Radome-Enclosed Antennas, Dennis J. Kozakoff

Analysis of Wire Antennas and Scatterers: Software and User's Manual, A. R. Djordjević, M. B. Bazdar, G. M. Bazdar, G. M. Vitosevic, T. K. Sarkar, and R. F. Harrington

Analysis Methods for Electromagnetic Wave Problems, E. Yamashita, editor

Antenna-Based Signal Processing Techniques for Radar Systems, Alfonso Farina

Antenna Design With Fiber Optics, A. Kumar

Broadband Patch Antennas, Jean-François Zürcher and Fred E. Gardiol

CAD of Aperture-Fed Microstrip Transmission Lines and Antennas: Software and User's Manual, Naftali Herscovici

CAD for Linear and Planar Antenna Arrays of Various Radiating Elements: Software and User's Manual, Miodrag Mikavica and Aleksandar Nešić

The CG-FFT Method: Application of Signal Processing Techniques to Electromagnetics, Manuel F. Cátedra, Rafael P. Torres, José Basterrechea, Emilio Gago

Electromagnetic Waves in Chiral and Bi-Isotropic Media, I.V. Lindell, S.A. Tretyakov, A.H. Sihvola, A. J. Viitanen

Fixed and Mobile Terminal Antennas, A. Kumar

Generalized Multipole Technique for Computational Electromagnetics, Cristian Hafner

Handbook of Antennas for EMC, Thereza Macnamara

Integral Equation Methods for Electromagnetics, N. Morita, N. Kumagai, and J. Mautz

IONOPROP: Ionospheric Propagation Assessment Program, Version 1.1: Software and User's Manual, Hernert V. Hitney

Four-Armed Spiral Antennas, Robert G. Corzine and Joseph A. Mosko

Introduction to Electromagnetic Wave Propagation, Paul Rohan

Introduction to the Uniform Geometrical Theory of Diffraction, D. A. McNamara

Microwave Cavity Antennas, A. Kumar and H. D. Hristov

Millimeter-Wave Microstrip and Printed Circuit Antennas, Prakash Bhartia

Mobile Antenna Systems, K. Fujimoto and J. R. James

Modern Methods of Reflector Antenna Analysis and Design, Craig Scott

Moment Methods in Antennas and Scattering, Robert C. Hansen, editor

Monopole Elements on Circular Ground Planes, M. M. Weiner *et al.*

Near-Field Antenna Measurements, D. Slater

Passive Optical Components for Optical Fiber Transmission, Norio Kashima

Phased Array Antenna Handbook, Robert J. Mailloux

Polariztion in Electromagnetic Systems, Warren Stutzman

Practical Phased-Array Antenna Systems, Eli Brookner *et al.*

Practical Simulation of Radar Antennas and Radomes, Herbert L. Hirsch and Douglas C. Grove

Radiowave Propagation and Antennas for Personal Communications, Kazimierz Siwiak

Reflector and Lens Antennas: Software User's Manual and Example Book, Carlyle J. Sletten, editor

Reflector and Lens Antenna Analysis and Design Using Personal Computers, Carlyle J. Sletten, editor

Shipboard Antennas, Second Edition, Preston Law

Small-Aperture Radio Direction-Finding, Herndon Jenkins

Spectral Domain Method in Electromagnetics, Craig Scott

Understanding Electromagnetic Scattering Using the Moment Method: A Practical Approach, Randy Bancroft

Waveguide Components for Antenna Feed Systems: Theory and CAD, J. Uher, J. Bornemann, and Uwe Rosenberg

For further information on these and other Artech House titles, contact:

Artech House
685 Canton Street
Norwood, MA 02062
617-769-9750
Fax: 617-769-6334
Telex: 951-659
e-mail: artech@artech-house.com

Artech House
Portland House - Stag Place
London SW1E 5XA England
+44 (0) 171-973-8077
Fax: +44 (0) 171-630-0166
Telex: 951-659
e-mail: artech@artech-house.com